208
Advances in Polymer Science

Advances in Polymer Science
Recently Published and Forthcoming Volumes

**Wax Crystal Control · Nanocomposites ·
Stimuli-Responsive Polymers**
Vol. 210, 2008

Functional Materials and Biomaterials
Vol. 209, 2007

**Phase-Separated Interpenetrating Polymer
Networks**
Authors: Lipatov, Y. S., Alekseeva, T. T.
Vol. 208, 2007

Hydrogen Bonded Polymers
Volume Editor: Binder, W.
Vol. 207, 2007

**Oligomers · Polymer Composites ·
Molecular Imprinting**
Vol. 206, 2007

Polysaccharides II
Volume Editor: Klemm, D.
Vol. 205, 2006

**Neodymium Based Ziegler Catalysts –
Fundamental Chemistry**
Volume Editor: Nuyken, O.
Vol. 204, 2006

Polymers for Regenerative Medicine
Volume Editor: Werner, C.
Vol. 203, 2006

Peptide Hybrid Polymers
Volume Editors: Klok, H.-A., Schlaad, H.
Vol. 202, 2006

**Supramolecular Polymers ·
Polymeric Betains · Oligomers**
Vol. 201, 2006

Ordered Polymeric Nanostructures at Surfaces
Volume Editor: Vancso, G. J., Reiter, G.
Vol. 200, 2006

Emissive Materials · Nanomaterials
Vol. 199, 2006

Surface-Initiated Polymerization II
Volume Editor: Jordan, R.
Vol. 198, 2006

Surface-Initiated Polymerization I
Volume Editor: Jordan, R.
Vol. 197, 2006

**Conformation-Dependent Design of Sequences
in Copolymers II**
Volume Editor: Khokhlov, A. R.
Vol. 196, 2006

**Conformation-Dependent Design of Sequences
in Copolymers I**
Volume Editor: Khokhlov, A. R.
Vol. 195, 2006

Enzyme-Catalyzed Synthesis of Polymers
Volume Editors: Kobayashi, S., Ritter, H.,
Kaplan, D.
Vol. 194, 2006

Polymer Therapeutics II
Polymers as Drugs, Conjugates and Gene
Delivery Systems
Volume Editors: Satchi-Fainaro, R., Duncan, R.
Vol. 193, 2006

Polymer Therapeutics I
Polymers as Drugs, Conjugates and Gene
Delivery Systems
Volume Editors: Satchi-Fainaro, R., Duncan, R.
Vol. 192, 2006

**Interphases and Mesophases in Polymer
Crystallization III**
Volume Editor: Allegra, G.
Vol. 191, 2005

Phase-Separated
Interpenetrating Polymer Networks

By Yuri S. Lipatov and Tatiana Alekseeva

 Springer

Prof. Dr. Yuri S. Lipatov
Institute of Macromolecular Chemistry
National Academy of Sciences of Ukraine
Kharkivske schausse 48
Kiev 02160
Ukraine
lipatov@i.kiev.ua

Tatiana T. Alekseeva
Institute of Macromolecular Chemistry
National Academy of Sciences of Ukraine
Kharkivske schausse 48
Kiev 02160
Ukraine
att7@yandex.ru

The series *Advances in Polymer Science* presents critical reviews of the present and future trends in polymer and biopolymer science including chemistry, physical chemistry, physics and material science. It is adressed to all scientists at universities and in industry who wish to keep abreast of advances in the topics covered.
As a rule, contributions are specially commissioned. The editors and publishers will, however, always be pleased to receive suggestions and supplementary information. Papers are accepted for *Advances in Polymer Science* in English.
In references *Advances in Polymer Science* is abbreviated *Adv Polym Sci* and is cited as a journal.

Springer WWW home page: springer.com
Visit the APS content at springerlink.com

Library of Congress Control Number: 2007928736

ISSN 0065-3195
ISBN 978-3-540-73071-2 Springer Berlin Heidelberg New York
DOI 10.1007/978-3-540-73552-6

Springer is a part of Springer Science+Business Media

springer.com

© Springer-Verlag Berlin Heidelberg 2007

Cover design: WMXDesign GmbH, Heidelberg
Typesetting and Production: LE-TEX Jelonek, Schmidt & Vöckler GbR, Leipzig

Printed on acid-free paper   02/3100 YL – 5 4 3 2 1 0

Advances in Polymer Science
Also Available Electronically

For all customers who have a standing order to Advances in Polymer Science, we offer the electronic version via SpringerLink free of charge. Please contact your librarian who can receive a password or free access to the full articles by registering at:

springerlink.com

If you do not have a subscription, you can still view the tables of contents of the volumes and the abstract of each article by going to the SpringerLink Homepage, clicking on "Browse by Online Libraries", then "Chemical Sciences", and finally choose Advances in Polymer Science.

You will find information about the

– Editorial Board
– Aims and Scope
– Instructions for Authors
– Sample Contribution

at springer.com using the search function.

Contents

1	**The Nature of IPNs**	6
1.1	Introduction	6
1.2	Definitions	8
1.3	Formation and Structure of Amorphous Polymer Networks	10

2	**Thermodynamics and Phase Separation in IPNs**	15
2.1	Thermodynamics of IPN Mixing	15
2.1.1	Phase Diagrams of Reacting Systems	16
2.2	Miscibility and Immiscibility of IPNs	21
2.3	Mechanisms of Phase Separation in IPNs	24
2.3.1	Nucleation and Growth	24
2.3.2	Spinodal Decomposition	35
2.3.3	Binder–Frisch Approach	44
2.3.4	Interfacial Region in IPNs	47
2.3.5	Composition and Ratio of Phases Evolved During Phase Separation	51
2.4	Nonequilibrium States of IPNs	52

3	**Heterogeneous Structure and Morphology of IPNs**	55
3.1	Cross-linking Density in IPNs	56
3.2	Free Volume	62
3.3	Parameters of Heterogeneous Structure of IPNs	70
3.4	Experimental Data on Heterogeneity	73
3.4.1	Sequential IPNs Based on Polyurethane and Styrene–Divinylbenzene Copolymer	74
3.4.2	Simultaneous IPNs Based on Polyurethane and Poly(urethane acrylate)	77
3.4.3	Sequential IPNs Obtained Using Anionic Polymerization	80
3.4.4	Microphase Structure of IPNs Based on PU and Ionomeric PU	84
3.4.5	Reaction Conditions and Microphase Structure	87
3.5	Characterization of IPN Structure via Small-Angle Neutron Scattering . . .	91
3.6	Features of IPN Morphology	96

4	**Relaxation Transitions and Viscoelasticity of IPNs**	103
4.1	Application of Mechanical Models	105
4.2	Viscoelastic Functions and Relaxation Transitions	111
4.3	Viscoelasticity and Curing Conditions	116
4.4	Properties of Thermoplastic Apparent IPNs	120
4.5	Features of Temperature Transitions	124
4.6	Contribution of an Interphase to Viscoelastic Properties	126
4.7	Effect of the Domain Sizes on the Glass Transition Temperatures	128

4.8   Relaxation Spectra of IPNs   130
4.9   Vibration-Damping Properties   137
4.10   Determining the Segregation Degree
        from Parameters of Relaxation Maxima   139
4.11   Dependence of Viscoelastic Properties on Segregation Degree   143

5   **Chemical Kinetics of IPN Formation and Phase Separation**   147
5.1   General Premises   147
5.2   Kinetic Characteristics of IPN Formation   148
5.3   Mutual Influence of the Constituent Networks on the Kinetics
        of IPN Formation   157
5.4   Relation Between Reaction Kinetics and Microphase Separation   164
5.4.1   Semi-IPN Based on Styrene–DVB Copolymer and PBMA   164
5.4.2   Semi-IPN Formed by Curing PU Network in the Presence of PBMA   167
5.4.3   IPNs Produced by Simultaneous Curing
        of Polyurethane and Polymerization of Butyl Methacrylate   169
5.4.4   Thermodynamics and Kinetics of Phase Separation   173
5.5   Reaction Kinetics and Properties of Evolved Phases   176
5.6   Kinetics of Formation of Sequential IPNs   179
5.7   Role of Kinetics in Formation of an Interphase and Segregation   181
5.8   Description of Reaction Kinetics in Terms of Phase Transition   183
5.9   Some Features of Rheokinetics and Structure States
        of Semi-IPNs During Their Formation   186
5.10   Special Features of Self-Organization During the Formation of IPNs   191
5.11   Kinetics of IPN Formation in the Presence of Filler   194

6   **Compatibilization in Phase-Separated IPNs**   199
6.1   General Premises   199
6.2   Factors with Influence on Determining IPN Compatibility   201
6.3   Compatibilization in IPNs Based on PU/PBMA   203
6.3.1   Introduction of Internetwork Grafting   204
6.3.2   Introduction of Compatibilizers   207
6.4   Compatibilization in IPNs Based on PU/PS   208
6.4.1   DMA Investigation   210
6.4.2   DSC Measurements   214

References   218

Author Index Volumes 1–209   229

Subject Index   233

Adv Polym Sci (2007) 208: 1–227
DOI 10.1007/12_2007_116
© Springer-Verlag Berlin Heidelberg
Published online: 12 September 2007

Phase-Separated Interpenetrating Polymer Networks

Yuri S. Lipatov (✉) · Tatiana T. Alekseeva (✉)

Institute of Macromolecular Chemistry, National Academy of Sciences of Ukraine,
Kharkivske schausse 48, Kiev 02160, Ukraine
lipatov@i.kiev.ua, att7@yandex.ru

1	The Nature of IPNs	6
1.1	Introduction	6
1.2	Definitions	8
1.3	Formation and Structure of Amorphous Polymer Networks	10
2	Thermodynamics and Phase Separation in IPNs	15
2.1	Thermodynamics of IPN Mixing	15
2.1.1	Phase Diagrams of Reacting Systems	16
2.2	Miscibility and Immiscibility of IPNs	21
2.3	Mechanisms of Phase Separation in IPNs	24
2.3.1	Nucleation and Growth	24
2.3.2	Spinodal Decomposition	35
2.3.3	Binder–Frisch Approach	44
2.3.4	Interfacial Region in IPNs	47
2.3.5	Composition and Ratio of Phases Evolved During Phase Separation	51
2.4	Nonequilibrium States of IPNs	52
3	Heterogeneous Structure and Morphology of IPNs	55
3.1	Cross-linking Density in IPNs	56
3.2	Free Volume	62
3.3	Parameters of Heterogeneous Structure of IPNs	70
3.4	Experimental Data on Heterogeneity	73
3.4.1	Sequential IPNs Based on Polyurethane and Styrene–Divinylbenzene Copolymer	74
3.4.2	Simultaneous IPNs Based on Polyurethane and Poly(urethane acrylate)	77
3.4.3	Sequential IPNs Obtained Using Anionic Polymerization	80
3.4.4	Microphase Structure of IPNs Based on PU and Ionomeric PU	84
3.4.5	Reaction Conditions and Microphase Structure	87
3.5	Characterization of IPN Structure via Small-Angle Neutron Scattering	91
3.6	Features of IPN Morphology	96
4	Relaxation Transitions and Viscoelasticity of IPNs	103
4.1	Application of Mechanical Models	105
4.2	Viscoelastic Functions and Relaxation Transitions	111
4.3	Viscoelasticity and Curing Conditions	116
4.4	Properties of Thermoplastic Apparent IPNs	120
4.5	Features of Temperature Transitions	124
4.6	Contribution of an Interphase to Viscoelastic Properties	126
4.7	Effect of the Domain Sizes on the Glass Transition Temperatures	128
4.8	Relaxation Spectra of IPNs	130

4.9   Vibration-Damping Properties   137
4.10   Determining the Segregation Degree
         from Parameters of Relaxation Maxima   139
4.11   Dependence of Viscoelastic Properties on Segregation Degree   143

5   **Chemical Kinetics of IPN Formation and Phase Separation**   147
5.1   General Premises   147
5.2   Kinetic Characteristics of IPN Formation   148
5.3   Mutual Influence of the Constituent Networks on the Kinetics
         of IPN Formation   157
5.4   Relation Between Reaction Kinetics and Microphase Separation   164
5.4.1   Semi-IPN Based on Styrene–DVB Copolymer and PBMA   164
5.4.2   Semi-IPN Formed by Curing PU Network in the Presence of PBMA   167
5.4.3   IPNs Produced by Simultaneous Curing
         of Polyurethane and Polymerization of Butyl Methacrylate   169
5.4.4   Thermodynamics and Kinetics of Phase Separation   173
5.5   Reaction Kinetics and Properties of Evolved Phases   176
5.6   Kinetics of Formation of Sequential IPNs   179
5.7   Role of Kinetics in Formation of an Interphase and Segregation   181
5.8   Description of Reaction Kinetics in Terms of Phase Transition   183
5.9   Some Features of Rheokinetics and Structure States
         of Semi-IPNs During Their Formation   186
5.10   Special Features of Self-Organization During the Formation of IPNs   191
5.11   Kinetics of IPN Formation in the Presence of Filler   194

6   **Compatibilization in Phase-Separated IPNs**   199
6.1   General Premises   199
6.2   Factors with Influence on Determining IPN Compatibility   201
6.3   Compatibilization in IPNs Based on PU/PBMA   203
6.3.1   Introduction of Internetwork Grafting   204
6.3.2   Introduction of Compatibilizers   207
6.4   Compatibilization in IPNs Based on PU/PS   208
6.4.1   DMA Investigation   210
6.4.2   DSC Measurements   214

References   218

Abstract The key physicochemical problems of the formation and properties of interpenetrating polymer networks (IPNs) are considered. The main feature that determines the structure and properties of IPNs consists of the thermodynamic incompatibility of two constituents that arises in the course of the chemical reactions leading to the formation of IPNs. The peculiarities of the chemical reactions of IPN formation are the dependence of the reaction rate of the formation of each network on the presence of another. The chemical reactions are accompanied by the processes of phase separation. The conditions of the phase separation (its rate and degree) are dependent on the chemical kinetics. The heterogeneous structure (two evolved phases and the interfacial region between them) and the thermophysical, viscoelastic, and other physical properties of phase-separated IPNs are governed by the degree of segregation of the system into two phases. The formation of IPNs proceeds under conditions of superposition of the chemical kinetics of two reactions and the physical kinetics of phase separation, both proceeding in nonequilibrium conditions. The result is incomplete phase separation and a lack of interpenetration over the entire volume of the system.

Keywords Interpenetrating polymer networks · Phase separation · Interfacial region · Segregation · Heterogeneous structure

Abbreviations

AA	Acrylic anhydride
AIBN	2,2-Azobisisobutyronitrile
AIPN	Apparent interpenetrating polymer network
ATR	Adiabatic temperature rise; attenuated total reflection
BA	Butyl acrylate
BACY	Bisphenol A based cyanate ester
BAG	Butylene adipate glycol
BMA	Butyl methacrylate
BMP	Bisphenol A based bismaleimide
CER	Cyanate ester resin
CF	Carbon filler
CPU	Crystallizable polyurethane
CRS	Creep rate spectroscopy
DAA	Diallyl adipate
DAP	Diallyl phthalate
DBTDL	Dibutyl tin dilaurate
DDM	Diaminodiphenylmethane
DEGAC	Diethylene glycol bis(allyl carbonate)
DGEBA	Diglycidyl ether of bisphenol A
DMA	Dynamic mechanical analysis
DMS	Dynamic mechanical spectroscopy
DMTA	Dynamic mechanical thermal analysis
DSC	Differential scanning calorimetry
DVB	Divinylbenzene
EGDMA	Ethylene glycol dimethacrylate
EPCN	Epoxypolycyanurate
FTIR	Fourier-transform infrared spectroscopy
HEMA	2-Hydroxyethyl methacrylate
HMDI	Hexamethylene diisocyanate
HTPB	Hydroxy-terminated polybutadiene
IGC	Inverse gel-permeation chromatography
IPN	Interpenetrating polymer network
IR	Infrared
LCST	Lower critical solution temperature
LPU	Linear segmented polyurethane
m-TMXDI	1,1,3,3-Tetramethylxylene diisocyanate
MADGEBA	Methacrylated diglycidyl ether of bisphenol A
MDI	Macrodiisocyanate
MEE	Monomethyl ester of ethylene glycol
Mg-HBA	Magnesium salt of p-hydroxybenzoic acid
MGP	α,ω-Dimethacryl-bis(triethylene glycol phthalate)
MM	Molecular mass
MMA	Methyl methacrylate

MMD	Molecular mass distribution
M_n	Number-average molecular mass
M_w	Weight-average molecular mass
MWS	Maxwell–Wagner–Sillars
NMDA	N-Methyldiethanolamine
NMR	Nuclear magnetic resonance
OEA	Oligoesteracrylate
OIG	Oligoisoprene hydrazide
OTMG	Oligotetramethylene glycol
OUA	Oligourethane acrylate
OUDM	Oligourethane dimethacrylate
PA	Polyacrylate
PBMA	Poly(butyl methacrylate)
PCU	Poly(carbonate urethane)
PDS	Polydeuterostyrene
PDMS	Poly(dimethylsiloxane)
PE	Polyester
PEA	Poly(ester acrylate)
PEI	Polyetherimide
PEO	Poly(ethylene oxide)
PES	Polyethersulfone
PHEMA	Poly(2-hydroxyethyl methacrylate)
PMA	Poly(methyl acrylate)
PMMA	Poly(methyl methacrylate)
PnBA	Phenyl butyl acetate
POPG	Poly(oxypropylene glycol)
PS	Polystyrene
PSn	Polysulfone
PU	Polyurethane
PUA	Poly(urethane acrylate)
Pur	Polyurea
PVA	Poly(vinyl acetate)
PVC	Poly(vinyl chloride)
PVME	Poly(vinyl methyl ether)
RIM	Reaction injection molding
SANS	Small-angle neutron scattering
SAXS	Small-angle X-ray scattering
SEM	Scanning electron microscopy
TBU	Triblock urethane
TDI	Toluene diisocyanate
TEM	Transmission electron microscopy
TGA	Thermogravimetric analysis
TMA	Trioxyethylene α,ω-dimethacrylate
TMDSC	Temperature-modulated differential scanning calorimetry
TMI	Benzene-1-(1-isocyanato-1-methyl ethyl)-3-(1-methylethenyl)
TMP	Trimethylolpropane
TPU	Thermoplastic polyurethane
TSC	Thermally stimulated conductivity
TSDC	Thermal stimulated depolarization current
UCST	Upper critical solution temperature

UP	Unsaturated polyester
UV	Ultraviolet
VER	Vinyl ester resin
WAXS	Wide-angle X-ray scattering
WLF	Williams–Landell–Ferry

Symbols

α	Conversion degree; segregation degree
β	Wave number
Γ	Gamma function
γ	Shear rate, interfacial tension
δ	Mechanical loss angle
$\Delta\eta$	Electron density fluctuation
η	Viscosity
θ	Scattering angle
λ	Wavelength; Takayanagi model parameter
λ_m	Optimal wavelength of spinodal decomposition
μ	Chemical potential
ν	Poisson coefficient; effective concentration of cross-links
ρ	Density; electron density
τ	Relaxation time; shear stress
φ	Volume fraction
χ	Flory–Huggins interaction parameter
ω	Circular frequency
ψ	Interaction parameter
B_{11}	Second virial coefficient
C	Concentration
D	Diffusion coefficient; domain size
$D(g)$	Debye function
E	Elastic modulus
E^*	Complex elastic modulus
E'	Real part of complex modulus
E''	Imaginary part of complex modulus
E_a	Activation energy
F	Helmholtz free energy
f	Initiator efficiency; fractional
f_c	Cross-link functionality
G	Gibbs free energy; shear modulus
g	Free energy
G'	Storage modulus
G''	Loss modulus
H	Enthalpy; function of relaxation times distribution; heat
$<h>$	Mean square end-to-end distance
ΔH	NMR second moment
I	Scattering intensity
$[I]$	Initiator concentration
K	Reaction rate constant
K_g	Propagation constant
K_t	Termination constant
$[kt]$	Catalyst concentration
l_C	Heterogeneity length

l_ρ	Average size of heterogeneity region
LA	Loss area
M	Molecular mass; diffusion mobility
M_c	Molecular mass between two cross-links
n	Reaction order
N_e	Number of repeating units between cross-links
P	Any property of a system
Q	Scattering vector
q	Wave vector
R	Amplification factor; gas constant
$<R>^2$	Mean square end-to-end distance
r	Domain radius
R_g	Gyration radius
R_p	Porod radius
S	Entropy; structural factor
S_ν	Specific interface surface
T_g	Glass transition temperature
T_2	Transverse relaxation time
t	Time
t_{mps}	Time of the onset of microphase separation
v	Molecular volume of repeating unit
V	Reaction rate
V_i	Initiation rate
V_G	Retention time
w	Weight fraction
W_{red}	Reduced reaction rate
Z	Coordination number

1
The Nature of IPNs

1.1
Introduction

The first monograph dedicated to interpenetrating polymer networks (IPNs) was published in 1979 in Russian; it was written by Y.S. Lipatov and L.M. Sergeeva. Within 2 years, the well-known monograph by L. Sperling, "Interpenetrating Polymer Networks and Related Materials", was published by Plenum Press [1]. Since that time more than 25 years has gone by. During this period of time, a great many works were published on IPNs and new concepts were developed; in addition, many aspects of IPN formation were studied and regularities established. This predetermined the necessity of a new analysis of the state of the problem, which is the aim of the present review. However, the field of IPNs is developed so extensively, that presently it is almost impossible in one review to consider all aspects of the synthesis, properties, and especially, application of new systems and new materials for their production. The main task here is to discuss the physicochemical features of IPN formation, including the reaction kinetics, kinetics of phase separation, and molecular properties, connected with processes that occur in IPNs during their synthesis etc.

In this review we deal only with phase-separated IPNs. There are some data about compatible IPNs, but a more typical and much more interesting situation arises when we consider superposition of chemical and physical processes, both having a mutual influence on one another.

After analyzing over many years the development of science, we have come to the conclusion that many up-to-date results often repeat what was already known, although accounting for the modernized or new concepts. We think we have no right to forget our predecessors, and hence in this review we give references to some rather old works, where the main principles were formulated.

First on the IPN ring, Dr. Jonas W. Aylsworth appeared in 1914, who, as we think now, was the first to have invented IPNs, not having been thinking about what the next generation of people would call what he invented. After long time, Millar (1960) [2] appeared, who proposed the definition of IPNs and supposed that two networks are really interpenetrating. It was nice but did not agree with thermodynamics. Then, almost simultaneously, came K. Frisch, H. Frisch, L. Sperling, and D. Klempner [3, 4] and works were published where all said (maybe in different words): no, these IPNs are phase-separated. What was left then from interpenetrating? No morphological data could give an answer but that in these systems, dual-phase morphology appears. Is this morphology the interpenetrating structure we search for? Again, generally, the answer was "no", as this morphology occurs only at a definite network ratio. The new development allowed us to say: yes, IPNs are two-phase systems (all were already sure it was true), only each phase evolved by phase separation may be considered as a true interpenetrating network, being in the state of forced compatibility, namely, in a quasi-equilibrium state.

Historical review of the development of IPNs was given by Sperling [1, 3, 4]. The first IPNs that attractd everybody's attention were synthesized by Millar in 1960 [2]. According to initial hypothesis, IPNs represent a very complex system consisting of two or more polymers, where different polymers are not chemically bonded but cannot be separated due to mechanical entanglements of chains created during the synthesis. So, it was assumed that in IPNs there exists molecular mixing of various polymers, which is the main factor determining their properties.

Since the concept of chemical topology was introduced [5], a number of investigators have prepared molecules exhibiting topological isomerism. Most of these molecules were catenanes, consisting of interlocking rings with no chemical bonds between them. Since 1970, attention has turned to IPNs [6–10]. If we consider a cross-linked polymer to be a linear molecule with macrocycles of various sizes along the chain, we can envision how cross-linking of the initially linear polymer in the presence of another cross-linked polymer can give rise to "polymeric catenanes" or IPNs. The definition of IPNs as catenanes was rather widespread in the literature after the work of H. Frisch. Since that time, IPNs whose synthesis was of a different nature were considered to be polycatenanes.

In reality, the situation with polycatenanes is much more complicated. The problem is that a catenane-like structure of IPN is hard to identify by any single method. The difficulty is that neither catenanes nor other topological compounds (knots, rotaxanes) differ from the mixture of their individual components [11, 12]. If IPNs were true polycatenanes, they would exhibit no phase separation. Therefore, catenane-like IPNs may be formed only in the case when the network components are miscible for the whole range of composition and temperature values [13], and when the IPNs have a one-phase homogeneous structure.

Presently, a great deal is known about several of various IPNs. The processing of IPNs was recently considered again by Frisch [14]. According to our opinion, there is no direct evidence that the IPNs under investigation were really polycatenanes.

Sperling [15] defines IPNs as a "combination of two polymers in network form, at least one of which is synthesized and/or cross-linked in the immediate presence of the other". This definition has nothing in common with catenanes or interpenetration between two components on a molecular level. Sperling relates IPNs to the new class of polymer blends, where network polymers are mixed. Really, the synthesis of IPNs is a new way of blending polymers, which cannot be mixed by traditional or thermoplastic methods due the lack of ability of cross-linked polymer to melt or to be dissolved in any solvent. Obtaining IPNs requires polymers of varying chemical nature. Combining them with known monomers and oligomers allows one to obtain very different materials on the basis of a relatively narrow set of initial components.

What is really interpenetrating in IPNs is the penetration of the problem connected with their formation. It is difficult to discuss kinetics without thermodynamics and vice versa. The thermodynamics of interaction between two network components and the reaction kinetics determine the onset of phase separation, the structure, and the viscoelasticity of IPNs. This is why we begin our consideration with thermodynamic analysis and the elucidation of the reaction kinetics and its interconnection with phase separation. Only after considering these problems does the possibility arise to analyze both the structure and the viscoelasticity of IPNs.

1.2
Definitions

The differences in the methods of synthesis, in morphology, and in thermodynamics etc. may be used as a basis for IPN classification. Following

Sperling [16], IPNs produced by different methods may be distinguished as follows:

(i) Sequential IPNs, where polymer network I is prepared first. Network I swells in monomer II and cross-linking agent and is then polymerized in situ. Thus, in sequential IPNs the synthesis of one network follows the synthesis of the other.

(ii) Simultaneous IPNs, where the monomers or prepolymers and cross-linking agents for synthesis of both networks are mixed together. The reactions are carried out simultaneously. It is important that the cross-linking reaction should proceed according to different mechanisms to avoid chemical interaction between macromolecules of two networks. Usually these mechanisms are polyaddition and radical polymerization (a very rare case is anionic polymerization) [17].

In the example described [18], both networks are formed through the mechanism of radical polymerization under conditions when different monomers and initiators of various activities are used to separate the processes of network formation.

In cases (i) and (ii), chain transfer via polymerization may take place and grafted IPNs may be formed [19]. These two IPN types are the principal ones since they characterize the very principle of the IPN synthesis. Other kinds of IPNs, considered by Sperling, are gradient IPNs, latex IPNs, and thermoplastic IPNs. Latex IPNs are formed from a mixture of two lattices, frequently exhibiting a "core" and "shell" structure. Cross-linking proceeds on the level of two latex particles, and although the mixture of two networks is really present, in this case it is very difficult to say anything about interpenetration because of the existence of shells, adsorption layers of surfactants, usually used to produce latexes, etc. Gradient IPNs are characterized by the existence of a composition gradient, which is a result of nonequilibrium swelling of a previously prepared network in monomers for another. Gradient IPNs are a particular case of sequential IPNs.

Another very important class of IPNs are semi-IPNs, namely, systems in which one of the components is a linear polymer. Semi-IPNs may be characterized as sequential or simultaneous IPNs depending on the way the linear polymer is introduced. It may play the role of the "host" polymer for the synthesis of sequential IPNs, or that of the "guest" polymer for the same sequential IPNs when introduced into the system by the swelling of a network in monomers forming a linear polymer. Semi-IPNs may also be obtained by a simultaneous method via mixing the initial components of the network with linear polymer and then curing the system. Semi-IPNs may be of two types depending on what polymer is formed first: linear or cross-linked.

It is worth noting that in spite of the principal difference between sequential and simultaneous methods of IPN formation, in many cases the true simultaneous IPNs are practically never formed because of the difference in the reaction rates of formation of the two networks. One network is usually formed earlier and serves as some kind of "host" for the second, "guest" network.

Various types of IPNs may also be classified by the mechanism of phase separation proceeding during IPN formation. These mechanisms are nucleation and growth, and spinodal decomposition. Differences in the conditions of phase separation predetermine the physical and morphological features of IPNs. As a rule, simultaneous IPNs are phase-separated by a spinodal mechanism, and sequential ones via a mechanism of nucleation and growth. Another approach to IPN nomenclature was proposed by Sperling [1], who used the differences in morphological features of IPNs.

From a morphological point of view, all IPNs may be conveniently divided into ideal, partly interpenetrating, and phase-separated. An ideal IPN is a system with a molecular level of mixing of constituent networks. Practically, as will be discussed below, it is impossible to obtain such a network due to the thermodynamic incompatibility of the constituent components. Partial IPNs are the result of incomplete mixing of components; however, they are characterized by one major relaxation maximum, often considered as a sign of compatibility. IPNs based on incompatible polymers are the most widespread and most studied. However, due to the thermodynamic reasons considered below, such IPNs may have very different properties and because of this their classification is very vague. It should also be noted that morphological classification does not allow one to distinguish the real level of mixing due to the different scale of electron microscopy measurements (not molecular scale). Besides, morphology gives no answer as to the phase composition of the structures revealed through electron microscopy.

Sperling has applied group theory to classify both polymer blends and IPNs [1]. The initial phases of IPN research had an irrational basis in the words of Lewis Carroll. Indeed, the first sentence for such systems was that they are interpenetrating. Only after many years of investigation was the use of the very term IPN finally justified. The term IPN was beautiful, interesting, and prosperous; it was an idea to yearn for. The real verdict was the result of an understanding of the structure and properties of the systems. In this review we shall try to show the development of various concepts and approaches relating to IPNs, both experimental and theoretical. IPNs represent one of the most rapidly growing areas in polymeric materials science.

1.3
Formation and Structure of Amorphous Polymer Networks

The development of concepts concerning the formation and structure of three-dimensional network polymers was for a long time exclusively based on chemical considerations. It was assumed that these systems are formed from monomers or oligomers containing two or more reactive groups capable of chemical interaction, or from macromolecules containing in the main chain some groups capable of chemical interaction with curing agents. According to such concepts, the network polymer should represent a gigantic macromolecule, in which all constituent chains are cross-linked by chemical bonds. These concepts had their origin in the network theory of rubber elasticity. However, theoretical calculations of the ultimate properties of cross-linked polymers (which should be determined by the strength of chemical bonds) did not confirm such a concept [20], as no essential difference was discovered in the strength between amorphous cross-linked and linear polymers. The network defects (cyclic formation and inactive chains situated between cross-links) [21] did not explain this either. Practically, no networks were found where the number of active chains could be determined from the functionality of the cross-links. For example, if the unit volume of a hypothetical network contains ν tetrafunctional knots, the number of active chains is $N_a = 2\nu$.

On the other hand, a chemical network alone cannot explain the variety of network properties. Physical intermolecular polar interactions, donor–acceptor interactions, etc. play a very important role in network formation. In network systems capable of forming physical cross-links, the contribution of such links to the effective network density may be very significant [22]. It is known that the whole group of modern polymer materials, thermoelastoplastics, contain only physical cross-links.

The statistical theory of network formation was developed by Flory, Stockmayer, and others [23–26]. The formation of IPNs obeys the general rules of polymer network formation, but is also much more complicated because the reaction is accompanied by phase separation. Thus, it is important to briefly review the concepts developed by Dušek [27–31]. According to Dušek, the network formation may be accompanied by phase separation of the reaction mixture (especially near the gel point), even if the initial reaction system was homogeneous. Dušek has formulated the conditions at which phase separation is possible in the course of three-dimensional polymerization, and derived equations describing phase equilibrium in such systems. The condition for phase separation is the equality between maximum network swelling in equilibrium with a solvent or the solvent–monomer mixture $(1 - V_C)$ and current swelling degree $(1 - V_C^0)$, i.e., $V_C = V_C^0$, where V_C and V_C^0 are volume fractions of the three-dimensional polymer. By increasing the cross-linking density, the equilibrium becomes distorted (it is evident that V_C^0 cannot be higher than V_C) and microseparation proceeds because of the immiscibility of the solvent with the formed polymer. The system instability may be the result of both the changes in interaction between the system components and the increasing cross-linking degree. The separation of the new phase in the network is observed as a macro- and microsyneresis when the evolution of the new phase leads to the formation of a dispersion. According to Dušek, the interaction may not necessarily lead to full instability. It is enough if some fluctuations of both the network segment and knot of the network arise in the system. From a thermodynamic point of view, the aggregation of coils is regulated by the effect of excluded volume and because of this the cross-linking may proceed nonuniformly.

Phase separation may also occur in curing systems without any solvent [32]. During curing of alkyd resins it was observed that starting from some conversion degree, the rate of viscosity increase in the system diminished whereas the polymerization degree increased. The authors conjectured that when a definite conversion degree in the curing system is reached, microgel particles of colloidal dimensions appear. In such microheterogeneous systems, low molecular products of reaction play the role of dispersion medium. The concentration of microparticles is low and their interaction is negligible. Because the reaction rate in heterogeneous medium is lower, the number of microparticles increases with their size remaining unchanged. After some degree of conversion, these particles flocculate and then interact chemically. Unfortunately, this interesting hypothesis was not experimentally confirmed.

A structural approach to polymer network formation was developed and proved experimentally by Lipatova [23, 33–35]. It is based on taking into consideration the strong intermolecular interaction between oligomeric components of the reaction system, which lead to the appearance of local ordered regions. In systems consisting of low molecular mass chains and branched molecules (before gel point), molecular aggregates with a definite degree of ordering may be formed (and were observed experimentally). Lipatova reached the conclusion that the reaction of cross-linking is not a reaction between isolated molecules, but between such ordered regions. The reaction of three-dimensional polymerization, or polyaddition and polycondensation, proceeds under microheterogeneous conditions. The preliminary ordering of reaction components makes it possible to cure the system under heterogeneous conditions as a result of micro- and macrophase separation in the course of the reaction.

The structural approach to network formation assumes that the physical characteristics of the reaction medium and the level of intermolecular interactions in the system are the determining factors in network formation. Strong intermolecular interactions and increasing viscosity on approaching the gel point lead to nonuniform distribution of cross-links in the system. It is interesting that after the gel point the reaction may be described by the Avrami–Erofeev equation [23, 33]. Parameters of the equation found experimentally show that after the gel point, the reaction is described by the first-order equation. The change of the reaction order and simultaneous liquid–solid transition and appearance of the interface (microheterogeneity) show that the reaction rate is determined by the structure formation rate, i.e., by the total area of microgel particles. Thus, cross-linking is the process of transition of the system from one state to another (liquid–solid) and is accompanied by increasing intermolecular interaction and formation of supramolecular structures. Because of this, in the course of the three-dimensional polymerization or polycondensation, and due to formation of highly branched macromolecules, microgel particles appear, which are nuclei of the network structure and whose formation leads to subsequent microphase separation of the system. The increase in the size of cross-linked microgel particles occurs either as a result of their direct interaction or through tying branched macromolecules [33].

The structural mechanism of network formation was developed further by Korolev [36]. According to Korolev, the heterogeneity of the curing system may arise not only as a result of phase separation, but also as a consequence of higher reaction rates in molecular aggregates, which lead, in turn, to acceleration of microsyneresis inside the aggregate. This effect enhances the aggregate localization in the reaction volume and, as a result, instead of a uniformly distributed macromolecular network, a polymer consisting of highly cross-linked grains of microscopic and submicroscopic dimensions with loose cross-linked interlayers between them is formed. It is worth noting that the concepts of the microheterogeneous structure of polymer networks were developed simultaneously with the network theory of rubber elasticity. In particular, Berlin [37] proposed the scheme of structural elements of the three-dimensional polymer, according to which such a polymer is comprised of network aggregates of limited size. In these aggregates the macromolecules are linked by chemical bonds, whereas the aggregates themselves are linked by both chemical and Van der Waals bonds.

The heterogeneity of amorphous polymer networks may be observed directly from small-angle X-ray scattering. The concepts discussed above play a very important role in the analysis of the structure and properties of IPNs. In the development of concepts describing the behavior of amorphous networks, a great role belongs also to various models of topological entanglements. The simplest case of an amorphous structure is a system where molecular chains are mutually nonintersecting. Many deviations from the kinetic theory of rubber elasticity may be explained by the existence of entanglements, which are formed in the reaction system before the gel point. In such systems the entanglements serve as additional cross-links. Mark and Tobolsky [38] have described a model of a cross-linked elastomer, in which the network knots are, at first, destroyed under the action of thermal fluctuations, and are restored later. In this case the typical viscoelastic behavior of an elastomer is observed: elastic response to fast loading and viscous flow and stress relaxation during loading. The problems connected with the networks of entanglements are considered in several reviews [39–41].

Frisch determines IPNs as topologically interpenetrating systems. According to Irzhak [25] the topological structure of a polymer is determined by the connectivity of the structural elements, and may be described as a graph independent of the real chemical and spatial structure and disposition of its elements in space. Topological knots are labile formations that reveal themselves at short-time tests or through equilibrium deformation in densely cross-linked networks.

As is noted in [24], in spite of the great importance of entanglements they were never observed experimentally. Deviations of the real elastic properties of the elastomer from those predicted by classical theory of rubber elasticity usually serve as evidence of the presence of entanglements. The molecular mass of the chain between entanglements is usually estimated from the value of the quasi-equilibrium modulus on the elasticity plateau [24, 42, 43]. There were several attempts to estimate the probability of entanglement formation from the geometric parameters of chains [41]. It was suggested that entanglement is intermolecular contact where one chain forms a loop around another. Calculations were done for the polymethylene chain under the condition that the loops are formed by four to six methylene groups, and by accounting for the geometry of the chains. In [44] the number of bonds between entanglements was found to be 175, which agrees with experimental data. However, in another case [45] it was found that a chain consisting of 700 links may form a loop only from the sequence consisting of at least 30 links. The calculation of the entanglement network density was performed by taking account of the Van der Waals molecular volumes and of the molecular packing coefficient. It was shown that the effective distance between loops should be such that the mass of this part of the chain exceeds the average molecular mass of the usual polymer [46]. It was also established that the formation of the loops is unfavorable from a thermodynamic point of view. The estimation of the probability of forming loops and cycles of more complicated configurations was also done [46]. The molecular entanglements may also be modeled as the chains in the entanglement tube (so-called reptation mechanism of the chain diffusion in highly concentrated systems) [47].

Unlike linear polymers, topological entanglements in cross-linked polymers affect not only dynamic, but also equilibrium mechanical properties. The equilibrium shear modulus at the temperature above glass transition consists of two components: σ_x estimated from the classical theory and σ_y connected with the network topology [29]. The share of topological component depends on the chemical cross-linking degree [48].

It may be suggested that, because it is not yet known whether IPNs follow either the lower (LCST) or upper critical solution temperature (UCST) phase diagram, the effect of temperature on miscibility cannot be established, since curing usually proceeds at elevated temperatures and estimation of miscibility is carried out at room temperature.

IPNs can be considered as a new class of polymer composite materials made of two different components. IPNs may also serve as hybrid binders for traditional composite materials. As distinct from traditional binders, hybrid binders [49] are the systems where segregation of microvolumes of constituent components has occurred due to incomplete microphase separation. The hybrid binders can be considered as self-reinforced (filled) or disperse-reinforced systems, where the size, distribution, and properties of the regions of microphase separation (quasi-particles of filler) are determined by the phase diagram of a binary or multicomponent system, conditions of system transition through the binodal and spinodal, and by the mechanism of phase separation. An essential feature of hybrid matrices is the fact that the microphase structure develops directly in the course of the chemical reaction of cross-linking and the formation of a network structure.

Now we would like to go back to the beginning to determine the relation between the "sentence" and the "verdict". From what has been said above, it follows that most IPNs are heterogeneous phase-separated materials where one can observe the interpenetration only on a submolecular level (dual-phase continuity [1]). The main factors controlling the extent of phase separation are [50, 51]: compatibility, or thermodynamic interaction between components, kinetics of reaction, composition, mobility of the polymer chain, and polymerization degree at the time of gelation. In such a way, it seems that a more correct sentence is either interpenetration on the submolecular level, or a microheterogeneous matrix with inclusions of one of the phases. What is left then of interpenetration? Quite a bit. As will be shown later, IPNs are not true thermodynamically equilibrated systems. However, each evolved phase may be considered as a quasi-equilibrium one, corresponding to the frozen state of miscibility (forced compatibility), which is characteristic of the preceding stages of IPN formation.

Thus, it is most probable that IPNs are two-phase microheterogeneous systems with some interpenetration between two constituent phases and with a molecular level of mixing in each phase, due to the impossibility for these phases to be separated under conditions of IPN formation.

2
Thermodynamics and Phase Separation in IPNs

2.1
Thermodynamics of IPN Mixing

The thermodynamics of mixing of two networks in IPNs is a determining factor affecting IPN formation and properties which, in turn, are determined by the processes of phase separation proceeding during IPN formation. These processes lead to the appearance of thermodynamic immiscibility of network components and to the development of a heterogeneous structure. This structure depends on the mechanism of phase separation during IPN formation. The present state of the problem of polymer mixing was recently reviewed [52]. Here we consider the main characteristics of mixing of two linear polymers or two networks, based on the general thermodynamic principles not considering the fundamentals.

2.1.1
Phase Diagrams of Reacting Systems

The microheterogeneous structure of IPNs arises as a result of the thermodynamic immiscibility (corresponding to the positive value of thermodynamic interaction parameter χ) that appears in the course of IPN formation and leads to the microphase separation of the system.

It is known that the thermodynamic stability of multicomponent systems is determined by the dependence of the Gibbs free energy on concentration. If $\partial^2 G/\partial\varphi^2 < 0$, the composition fluctuations in the system increase, and the system separates into two phases with equilibrium composition (G is the free energy, φ is the volume fraction of one of the components of the binary system). The best way to characterize the binary system is via its phase diagram. However, for IPNs this task is practically impossible because phase separation proceeds simultaneously with IPN formation, so that the composition and the molecular mass of network fragments changes constantly, thus creating thermodynamic immiscibility.

In spite of this, an attempt has been made to construct a phase diagram for the curing semi-IPN based on poly(butyl methacrylate) (PBMA) and styrene–divinylbenzene (DVB) copolymer (Fig. 1) [53]. The system polystyrene (PS)/PBMA shows limited compatibility at low concentrations and low molecular mass of components, and full incompatibility at higher concentrations and molecular weights. With increasing conversion, the quality of the "solvent" (mixture of styrene and DVB) becomes poorer in relation to PBMA. As a result, after reaching a certain degree of conversion, the system can no

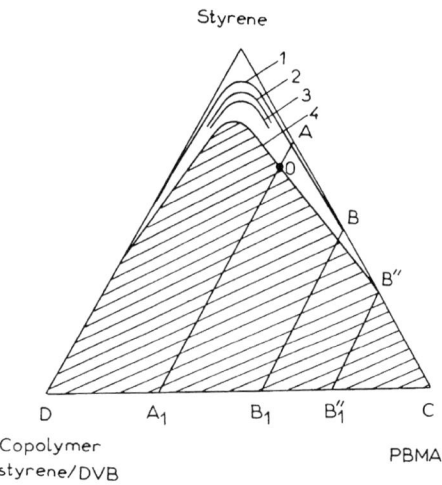

Fig. 1 Phase diagram of semi-IPN composed of styrene–DVB–PBMA at (1) 333 K, (2) 343 K, (3) 353 K, (4) 363 K [53]

longer preserve phase uniformity. It begins to separate into two phases. Thermodynamic incompatibility appears at a very low conversion degree. This fact is in good accord with the phase diagram of the semi-IPN obtained at various temperatures (Fig. 1). The region corresponding to the two-phase semi-IPN state (hatched region) is much larger than the regions of the one-phase state. In the phase diagram, heterogeneous regions are separated from the homogeneous ones by the binodal curve. In the course of semi-IPN formation, the polymerizing system passes from point A to point A_1, corresponding to the initial mixture and to the IPN, respectively. After the border of the two-phase system (point 0) is reached, phase separation can occur. Increasing the reaction temperature does not change the shape of the diagram. However, the area of the one-phase state slightly increases. Usually, for the ternary systems with UCST, the increase in temperature broadens the region of the one-phase state. If one analyzes the region of IPN compositions situated inside the B'' B_1'' C triangle, one can see that in this region the one-phase state cannot be realized at all. The B''–B_1'' line corresponds to a PBMA content of 70 mass %. Therefore, polymerization of the mixture in which the PBMA content exceeds these values (at the corresponding temperature) is accompanied by phase separation from the very beginning, i.e., when the content of copolymer is still insignificant. From the phase diagram one can also see that the copolymer/PBMA system in the absence of "solvent" is incompatible for the whole range of PBMA concentrations (C-D line) due to the high molecular mass (MM) of PBMA (500 000).

Recently, phase diagrams of ternary IPNs have been obtained [54]. The semi-IPNs based on cross-linked polyurethane (PU) and linear poly(vinyl acetate) (PVA) have been investigated by adding the third component, PBMA. The phase diagram was constructed using the Flory–Huggins theory for

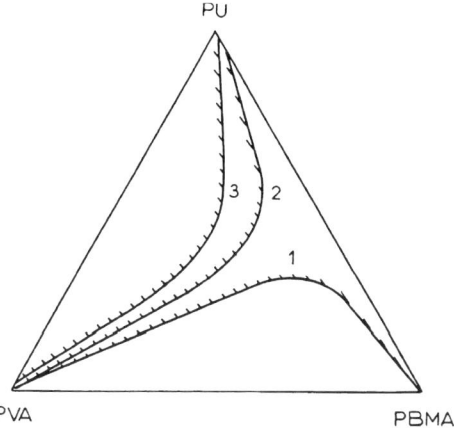

Fig. 2 Phase diagram of the ternary system PVA–PBMA–PU at (1) 384 K, (2) 400 K, (3) 476 K [54]

ternary systems, which accounts for interaction parameters between each pair of components present in the system. The phase diagram is presented in Fig. 2. It can be seen that introduction of the third component (PBMA) leads to the appearance of a region of component compatibility. With increasing temperature, this region broadens. These results are connected to the role of asymmetry of interaction in compatibility. In this case, the interaction parameter is expressed according to Eq. 1 [55]:

$$\chi = \chi_{12}\varphi_1\varphi_2 + \chi_{13}\varphi_1\varphi_3 + \chi_{23}\varphi_2\varphi_3 , \tag{1}$$

where χ are the interaction parameters between two species and φ are the respective volume fractions. If components 2 and 3 are incompatible ($\chi_{23} > 0$) (PVA–PU), then by introducing PBMA, which is compatible with both PVA and PU (χ_{12} and $\chi_{13} < 0$), the total interaction parameter χ may become negative and the region of compatibility is created (Fig. 2, curve 1). With increasing temperature, the values of χ_{12} and χ_{13} become more negative and their contribution into χ increases. As a result, χ becomes more negative and thermodynamic compatibility increases (Fig. 2, curves 2, 3). It is worth noting that, in this case, the third component was introduced into the curing system and may have influenced the kinetics of the reaction.

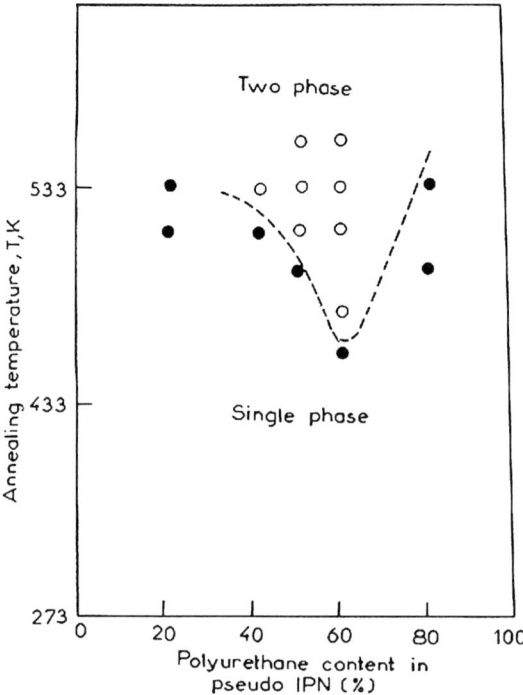

Fig. 3 Phase diagram of a semi-IPN composed of PU–PVC, showing the LCST [4]

The composition–temperature diagram for the semi-IPN *net*-polyurethane-*inter*-poly(vinyl chloride) (PU–PVC) has been obtained (Fig. 3) [4]. For this system, the LCST was found at approximately 393 K. Below this temperature the system is one-phase and above this temperature is two-phase. Very interesting results have been obtained by Sperling for the simultaneous IPNs based on PU–poly(methyl methacrylate) (PMMA) (Fig. 4) [4]. The metastable phase diagram has the form of a tetrahedron. Four vertices represent four pure components: two polymers, PMMA and PU, and two monomers, methyl methacrylate (MMA) and "U". The MMA–PMMA reaction line is fundamentally different from the "U"–PU one. Four triangular faces of the tetrahedron represent ternary systems. In the majority of the reaction, the PMMA is gelled after about 0.8% conversion, indicated by the plane G_1-U-PU. The phase separation curve for the ternary system MMA–PMMA–"U", the front left of the triangle, is represented by points C-D-A-E. Similarly, the phase separation curve for the MMA–PMMA–PU system, the rear triangle, is indicated by points J-B-K-L. The entire tetrahedron volume is divided into two regions: one phase-separated and one single-phased, represented by points C-D-A-E-L-K-B-J. The author notes the "sail-like" shape of this surface. Consider the curve A-B, which is the intersection of the PMMA gelation plane with that of the phase separation sail-like surface. This is a critical curve represent-

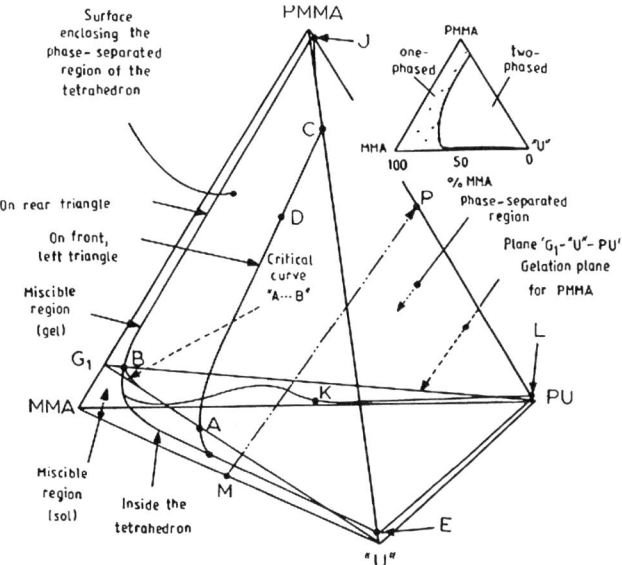

Fig. 4 Tetrahedron for the simultaneous IPN from PU and PMMA, showing the gelation plane of the PMMA and the curvilinear plane of phase separation. *Inset*: experimental data for the phase separating part of the front left triangle, showing the appearance of turbidity during MMA polymerization as an indication of phase separation [4]

ing simultaneous PMMA gelation and phase separation of the PU from the PMMA. Reactions moving to the left of this curve (as shown) will have the PMMA gel before phase separation, while reactions to the right of A-B will phase separate before gelation. The author emphasizes that for such a system the reaction may proceed in many possible directions. The key is the order in which the reacting system meets the three eventual surfaces inside the tetrahedron. This work by Sperling is of great importance for understanding the formation of simultaneous IPNs with various properties and is, in our opinion, the most detailed phase diagram of polymerizing IPNs.

In other work [55–59] a partly complete phase triangle diagram of the polymerization of styrene plus cross-linker in the presence of cross-linked polybutadiene has been constructed based on the data of transmission electron microscopy (TEM). This diagram shows the intercept with the binodal and spinodal curves at constant temperature.

The polymerization kinetics of simultaneous IPNs based on poly(n-butyl acrylate) and epoxy resins in situ have been studied by Fourier-transform infrared (FTIR) spectroscopy [60]. Three critical events occurred during the polymerization, namely, gelation of polymer network I, gelation of polymer network II, and phase separation of one polymer from the other. For these systems metastable phase diagrams describing the relation between the three events were constructed. Three-dimensional tetrahedrons characterizing the four-component system (two monomers and two polymers) allowed the visualization of these three events and also defined some critical points, for example, the loci of the points where simultaneous gelation of the two networks occurs. A triple critical point was identified, where both polymer gel and phase separate from each other simultaneously. The inside of the tetrahedrons was investigated using partially reacted model compounds.

Phase diagrams for semi-IPNs on poly(ethylene oxide) (PEO)/PMMA were also found [61]. The PEO network was produced by acid-catalyzed self-condensation of α,ω-bis(triethoxysilane)-terminated PEO in the presence of a small amount of water. The PMMA was formed by radical polymerization in the presence of DVB as cross-linker. The reaction conditions were adjusted to obtain similar cross-linking kinetics for both reactions. The phase diagrams were constructed by measuring the composition of the IPN at the moment of the appearance of phase separation, as indicated by the turbidity appearance. This composition could be determined because the siloxane cross-links of the PEO network could be hydrolyzed in aqueous NaOH with the formation of linear, soluble PEO chains. These phase diagrams were compared with phase diagrams of blends of linear polymers and semi-IPNs (cross-linked PMMA and linear PEO) obtained under similar conditions, i.e., polymerization of MMA in the presence of varying amounts of PEO. It was observed that the form of the phase diagrams of the linear polymers is similar to that of the IPNs, but is quite different from that of the semi-IPNs. The authors presented the ternary phase diagrams of MMA/linear PMMA/linear PEO and MMA/cross-linked PMMA/linear PEO systems at different temperatures. It was found that these systems have smaller miscibilities compared with blends, since the heterogeneous regions of the phase diagrams are larger than the one-phase regions. The area of the homogeneous region increases slightly with temperature. The decreasing thermodynamic miscibility in semi-IPNs could be ascribed to the unfavorable contribution of elastic stretching of the initially formed PMMA to the Gibbs free energy [62]. The authors do not believe that the degree of phase separation should depend on the cross-linking density. Really, at high density no phase separation becomes possible and the system is "chemically quenched", being in the state of forced compatibility. The latter term was introduced [63] to describe the role of mutual entanglements in preventing phase separation.

The comparison of the phase behavior of semi-IPNs based on almost miscible polymers—linear PS and poly-α-methylstyrene cross-linked by DVB and mixtures of the corresponding homopolymers—has shown that for homopolymer blends only one glass transition is observed, its position obeying the Fox equation. Simultaneously, for semi-IPNs there were two glass transitions, far from the glass transition temperatures of the components, for the same compositions where linear blends are miscible. A difference in phase behavior between blends and semi-IPNs seems to be evident. However, no phase diagrams allow one to determine the mechanism of phase separation.

2.2
Miscibility and Immiscibility of IPNs

When considering the formation of IPNs the first question to ask is whether both networks are compatible. The thermodynamic criterion of miscibility is the negative value of the Gibbs free energy ΔG (or of the enthalpy, ΔH). For the first time the enthalpy of two networks mixing was measured [64, 65] using inverse gas chromatography [66]. In the following expression, ΔH_i is the excess enthalpy of mixing the sorbate with the i-th networks, ΔH_{mix} is the excess enthalpy of mixing the sorbate with IPNs, $\Delta H_{1,2}$ is the enthalpy of mixing of the two networks, and w_i is the fraction of one of the networks:

$$\Delta H_{mix} = \sum w_i \Delta H_i - \Delta H_{1,2} . \tag{2}$$

The calculations were done for IPNs based on PU and styrene–DVB copolymer.

Calculations have shown that in almost all cases the values of $\Delta H_{1,2}$ are positive. It is known that positive values of the enthalpy of mixing show that such a pair is thermodynamically incompatible. Thus, in spite of forming the monolithic system, two constituent networks are not compatible, and formation of IPNs as a whole is the result of synthesis in the regime of forced compatibility. Only in the regions where the amount of one of the network

components is small may the system be compatible. To study the nonequilibrium state one needs to estimate the free energy of mixing of the constituent
networks. For this purpose the method of vapor sorption was employed [67].

From experimental data on benzene vapor sorption by all polymer systems, the changes in partial free energy of benzene by sorption (dissolution)
may be found as

$$\Delta\mu_1 = \frac{1}{M}RT\ln\frac{P}{P_0}, \tag{3}$$

where M is the molecular mass of benzene and P/P_0 is the relative vapor pressure. The value of $\Delta\mu_1$ changes with solution concentration from 0 to ∞. To
calculate the free energy of mixing of the polymer components with solvent
we need to know the changes in partial free energies of the polymers (individual networks, traditional IPNs, and layers of gradient IPNs). $\Delta\mu_2$ is the
difference between the polymer chemical potential in a solution of a given
concentration and in pure polymer under the same conditions.

$\Delta\mu_2$ for the polymer components has been calculated according to the
Gibbs–Duhem equation,

$$u_1\frac{d\left(\Delta\mu_1\right)}{du_1} + u_2\frac{d\left(\Delta\mu_2\right)}{du_2} = 0, \tag{4}$$

where u_1 and u_2 are the mass fractions of solvent and polymer. Hence,

$$\int d\left(\Delta\mu_2\right) = \int \frac{u_1}{u_2}d\left(\Delta\mu_1\right), \tag{5}$$

and by integrating over definite limits the values of $\Delta\mu_2$ may be found from
the experimental data.

The average free energy of mixing of solvent with the individual components, with IPNs of various compositions, and with gradient IPN layer for
solutions of different concentrations was estimated according to the equation:

$$\Delta g^m = u_1\Delta\mu_1 + u_2\Delta\mu_2. \tag{6}$$

Figure 5 shows the change of the free energy of mixing of benzene with
individual networks and with IPNs. All the systems studied (PU–benzene,
copolymer–benzene, gradient IPN–benzene) are thermodynamically stable
systems, as for them the following condition is valid:

$$\frac{\partial^2\Delta g^m}{\partial u^2} > 0. \tag{7}$$

From the concentration dependence of the free energy of mixing of the solvent and the system components, using the Hess law and thermodynamic
cycles [68], we have calculated the changes in free energy between two constituent networks.

The change in Gibbs free energy ∂g^* was found for an IPN based on PU
and poly(urethane acrylate) (PUA) [69] from the data on the free energy of

mixing of IPNs and initial networks with the solvent by measuring vapor sorption:

$$\partial g^* = - w_1 \Delta G_1 - w_2 \Delta G_2 + \Delta G_3 , \tag{8}$$

where ΔG_1, ΔG_2, and ΔG_3 are respectively the free energies of mixing of the first and second networks and IPNs with vapor, and w_1 and w_2 are the mass fractions of networks. The results are given in Fig. 5. It can be seen that the free energy of mixing of two networks is positive in the whole composition range, showing incompatibility of components. When performing such calculations, one has to bear in mind that the separate network and the same network in the IPN are not the same, and the results may be different due to this circumstance. As one can see, the phase diagram has a bimodal shape. The minimum may be connected to the formation of the transition zone (see below).

The initial stage of IPN formation may be considered as a thermodynamic equilibrium process (mixing monomers with oligomers or swelling of the matrix network in components of the second one). However, the process of separation proceeds very slowly due to the high viscosity of the system, as well as due to entanglements between chains. From what was said above it follows that the concept of topological networks mixed on the molecular level is only an abstraction, as well as the concept about ideal networks in general.

The data on the sorption isotherm allow us to calculate the affinity parameters in the IPN–solvent system (namely, chemical potentials of IPN and solvent, $\Delta \mu_2$ and $\Delta \mu_1$, and enthalpies and entropies of swelling). It was shown that introduction of the second network increases the thermodynamic

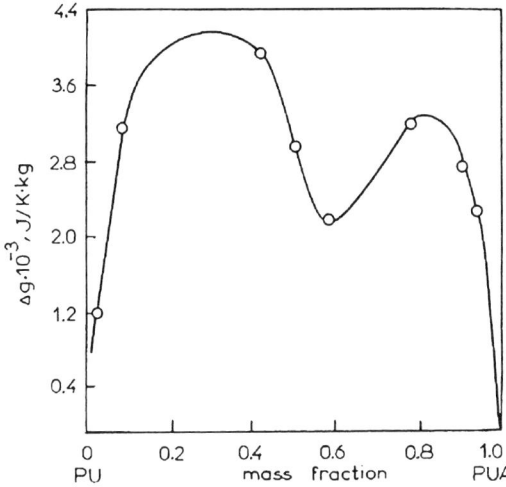

Fig. 5 Change in the free energy of the network mixing as a function of composition for an IPN based on PU and PUA [69]

stability of the IPN–solvent system, i.e., the IPN has a greater affinity to the solvent as compared to each network separately. From chemical potentials determined at various temperatures, one can compute the change in the partial specific enthalpy and entropy during sorption [69]:

$$\Delta H_2 = \frac{T_1 \partial \mu_2 (T_2) - T_2 \partial \mu_2 (T_1)}{T_1 - T_2} . \tag{9}$$

It was observed that the enthalpy of mixing of IPN and pure network with the solvent is positive. From the data on the concentration dependence of the partial specific entropy of the IPN and of constituent networks, a conclusion about the change in chain flexibility can be drawn: the higher is the value of $T\Delta S_2$, the higher is the flexibility. The entropy of IPNs is higher than that of a pure network and increases with the increasing fraction of the second network. This effect is the result of heterogeneity of the IPN structure and is connected with the formation of a transition zone with lower density packing and with directivity of the IPN as compared with initial networks. Similar results were obtained for semi-IPNs based on PU and poly(vinyl pyrrolidone) [70]. From the data on vapor sorption the free energy of mixing Δg^m of components with the solvent has been calculated, and using thermodynamic cycles the change in the free energy of mixing between two components was calculated as a function of IPN composition. In all cases the free energy of mixing was positive and the components could be considered as immiscible.

Therefore, all the IPNs are thermodynamically unstable. The peculiarity of IPN formation determines their morphology, which is formed in the course of phase separation by curing. Below we discuss two possible mechanisms of phase separation in IPNs.

2.3
Mechanisms of Phase Separation in IPNs

Two mechanisms of phase separation are nucleation and growth and spinodal decomposition. The theory of spinodal decomposition was developed by Cahn and Hilliard [71]. In the region of the phase diagram inside the spinodal, phase separation leads to the formation of microregions with compositions that deviate from the system composition by a very small amount.

2.3.1
Nucleation and Growth

Morphological data showing that IPNs are highly heterogeneous structures allowed Sperling et al. [72–74] to propose the mechanism of IPN formation called nucleation and growth. A thermodynamic theory of IPN morphology has been developed in [74]. Its main assumption is that there are two separate states of polymer I and polymer II separated one from another. In state 2, the network I is positioned in a uniformly swollen network in polymer II, if both polymers form regular solutions. At the end of reaction, phase separation proceeds with the appearance of spherical domains of polymer II in matrix I without any volume changes. It is assumed that each state corresponds to the state of thermodynamic equilibrium. This theory is rather complicated and it may only be applied to sequential IPNs. However, it is possible that the mechanism of phase separation really depends on the type of IPN (simultaneous or sequential).

Sperling has studied theoretical conditions for the formation of domains in sequential IPNs using cross-linking degree for each network, as well as thermodynamics of mixing and interfacial tension for sequential IPNs, where separation occurs by the nucleation mechanism. The derivation of the basic equation for IPN domain diameters is based on a physical model of sequential IPNs, according to which polymer II, which is formed in a swollen network I, constitutes a spherical core and is in a contracted (deformed) state, while polymer I surrounds the core and is in an expanded (deformed) state.

Several assumptions were made for this derivation: (1) a thermodynamic equilibrium process exists throughout the development of the domain formation; (2) the domains are spheres with identical diameters; (3) the polymer networks obey Gaussian statistics; and (4) a sharp interfacial boundary exists between the two phases (we have to note that the validity of these assumptions is rather questionable). The authors present the process of domain formation in the following way (Fig. 6) [75].

Initially, in state 1, network I is completely separated from monomer II. In state 2, the polymer network is swollen with the monomer mixture II. The path from state 1 to state 2 is accompanied by the mixing of polymer I and

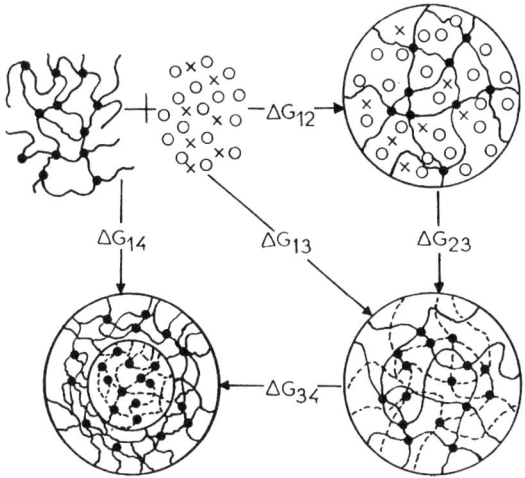

Fig. 6 Simplified path of domain formation [75]

monomer II, and by mutual concomitant expansion of polymer I caused by swelling with the monomer II mixture. The free energy of polymerization during the transfer from state 2 to state 3 is ignored. Also, the enthalpy 1, 2 contact energies between monomer II and polymer I are assumed to be the same as the polymer II–polymer I enthalpy contact energies. State 3 is the hypothetical, mutually mixed state, where polymers I, II are mixed and mutually diluted. Demixing (phase separation) between polymer I and polymer II with concomitant deformation of polymer II and further deformation of polymer I into a shell leads to state 4. State 4 is a phase-separated state with a spherical domain of polymer II surrounded by polymer I deformed into a spherical shell.

Referring to Fig. 6, the molecular rearrangements taking place during the transformation between state 3 and state 4 require amplification. As the authors remark, no covalent bond is broken during the process, as would be required in the case of literal interpretation of the model. Instead, phase separation ensues at an early stage of polymerization of monomer II, when the free energy of mixing becomes positive and the second derivative of the free energy of mixing with respect to composition is negative. This probably happens at or before the gel point. Thus, the molecular migration begins earlier (which is illustrated by the model), and hence stage 3 is unlikely.

Early research on the thermodynamics of the domain formation was performed by Donatelli, Sperling, and Thomas [74], who considered the free energy changes connected with surface effects and elastic free energy changes due to swelling.

Consider, following Yeo, Sperling, and Thomas [73], the thermodynamics of the process. For a closed system at constant pressure and temperature, the Gibbs free energy is given by

$$\Delta G = \sum \Delta H_{i,i+1} - T \sum \Delta S_{i,i+1} \quad (i = 1, 2, 3), \tag{10}$$

where $\Delta H_{i,i+1}$, $\Delta S_{i,i+1}$ represent the enthalpy and entropy changes involved in the process of transition from state i to state $i + 1$, respectively. Extending Eq. 10 to the domain formation process, the free energy change for polymer II domain formation, ΔG_d, can be expressed as

$$\Delta G_d = \sum_{i=1}^{3} \Delta H_{i,i+1} - T \sum_{i=1}^{3} \Delta S_{i,i+j} + \Delta G_i, \tag{11}$$

where ΔG_i represents the interfacial free energy change for domain formation and corresponds to transition from state 3 to state 4. From Fig. 6, we see that the path from state 1 to state 3 via state 2 can be replaced by the direct path from state 1 to state 3. In this case the following equation is valid:

$$\Delta G_d = \Delta H_{13} + \Delta H_{34} - T \left(\Delta S_{13} + \Delta S_{34} \right) + \Delta G_i. \tag{12}$$

In fact, ΔH_{13} and ΔH_{34} are the heat of mixing and the heat of demixing between polymers I and II, so that the sum of these two terms can be assumed to be zero. The quantity ΔS_{13} is equal to the sum of the entropy of mixing ΔS_m and the entropy change for the elastic deformation of polymer I being swollen with polymer II, ΔS^I_{sw}. The quantity ΔS_{34} is equal to the sum of demixing entropy change ΔS_{dm}, the rearrangement entropy change for elastic deformation of polymer II network upon deswelling ΔS^{II}_{dsw}, and the entropy change for elastic deformation of polymer I network ΔS^I_{df}. Here, ΔS_m and ΔS_{dm} cancel each other. The interfacial free energy change ΔG_i consists of the interfacial free energy change for domain formation, ΔG^0_i. A term for the entropy change on placing polymer I, II molecules in each domain, ΔS_p, must also be added. In summary, the free energy change for polymer II domain formation is expressed in the following way:

$$\Delta G_d = T \left(\Delta S^I_{sw} + \Delta S^I_p + \Delta S^{II}_{dsw} + \Delta S^I_{df} \right) + \Delta G^0_i . \tag{13}$$

On the basis of the proposed model, the authors derived expressions for all terms in Eq. 13. Equations 14–17 give these expressions.

$$\Delta S^I_{sw} = \frac{\pi}{12} \left(\frac{\varphi_1}{\varphi_2} \right) \nu_1 R \left(3\varphi_1^{-2/3} - 3 + \ln \varphi_1 \right) D_2^3 , \tag{14}$$

$$\Delta S_p = -\frac{\pi}{6} R \left(\frac{\varphi_1 \rho_1}{\varphi_2 M_1} \ln \varphi_1 + \frac{\rho_2}{M_2} \ln \varphi_2 \right) D_2^3 , \tag{15}$$

$$\Delta S^{II}_{dsw} = -\frac{\pi}{12} \nu_2 R \left(3\varphi_2^{2/3} - 3 - \ln \varphi_2 \right) D_2^3 , \tag{16}$$

$$\Delta S^I_{df} = \frac{\pi}{4} \nu_1 R \left(1/\varphi_2 \right) \left(2\varphi_1^{1/3} \right) \left(1/\varphi_2 \right) \left(2\varphi_1^{1/3} - \varphi_1^{4/3} - \varphi_1 \right) D_2^3 . \tag{17}$$

The interfacial free energy is proportional to the intrinsic interfacial tension between the two polymers, γ^0, and is equal to $\Delta G^0_i = \pi \gamma^0 D_2^2$ for spherical domains. Inserting all these quantities in Eq. 12 and taking the first partial derivative with respect to domain size D_2, $\partial(\Delta G_d)/\partial D_2$, the following expression for D_2 was derived:

$$D_2 = 4\gamma^0 \left[RT \left(A\nu_1 + B\nu_2 C \right) \right]^{-1} , \tag{18}$$

where the values A, B, and C are given by

$$A = (1/2)(1/\varphi_2) \left(3\varphi_1^{1/3} - 3\varphi_1^{4/3} - \varphi_1 \ln \varphi_1 \right) , \tag{19}$$

$$B = (1/2) \left(\ln \varphi_2 - 3\varphi_2^{1/2} + 3 \right) , \tag{20}$$

$$C = \frac{\varphi_1}{\varphi_2} \frac{\rho_1}{M_1} \ln \varphi_2 + \frac{\rho_2 \ln \varphi_2}{M_2} . \tag{21}$$

In these equations, φ are the volume fractions of polymers, ρ are polymer densities, ν is the number of moles of effective network chains, and M are molecular masses (which go to infinity for full IPNs). In the same work [73] the authors compared theoretical predictions with experimental data for IPNs based on cross-linked phenyl butyl acetate (PnBA) and PS. In spite of many arbitrary assumptions (including the one stating that a spinodal decomposition process rather than binodal phase separation actually takes place), rather good agreement between calculated and experimental domain diameters for various compositions of IPN and various cross-linking levels was found. It was pointed out that the domain size is very sensitive to the value of interfacial tension [74].

To describe the formation of the system in the course of reaction another model was proposed [76], which allowed prediction of the fraction, the composition, and the average radius of the disperse phase (domains) evolved during reaction. The model is based on the application of the Flory–Huggins equation, in which the entropy term is a decreasing function of the conversion degree due to an increase in the polymerization degree. As a result, phase separation occurs at a definite stage of reaction. On the basis of this model, equations describing nucleation and growth rate (coalescence) have been derived. The model predicts the possibility of phase inversion, of slowing down of the separation at the gel point, and of the increase of the fraction of domains and their size as functions of the curing temperature. Later [77] an expanded version of the model was proposed allowing the domain dimension distribution to be calculated and the factors determining this distribution to be established. The model was developed for the epoxy resin–rubber system (statistical copolymer of butadiene with acrylonitrile with end carboxylic groups–cycloaliphatic epoxy resin with amine hardener). Such a system is characterized by the evolution of an elastomeric phase with bimodal distribution of particles. Suppose that rubber component 2 is dissolved in epoxy solvent 1. Then the Flory–Huggins equation will have the following form:

$$\Delta S_v = \left(\frac{RT}{\overline{V}_{10}} \right) \left[\left(\frac{\varphi_1}{z_1} \right) \ln \varphi_1 + \left(\frac{\varphi_2}{z_2} \right) \ln \varphi_2 + \chi \varphi_1 \varphi_2 \right] , \tag{22}$$

where φ_1 and φ_2 are volume fractions of components, $\overline{V}_{10}$ is the initial volume of the network defined as

$$\overline{V}_{10} = \overline{M}_{n,10}/\rho_1 , \tag{23}$$

where $\overline{M}_{n,10}$ is the initial number-average molecular weight and ρ_1 is the density. Let ς = (epoxy equivalents)/(amine equivalents) = $2B_2/4A_4$, where B_2 and A_4 are numbers of moles of each monomer component. Then

$$\overline{M}_{n,10} = \left(M_{A_4} \varsigma M_{B_2} \right) / \left(1 + 2\varsigma \right) . \tag{24}$$

For the stoichiometric mixture of components $\varsigma = 1$. If z_1 and z_2 are ratios of molar volumes of both components over V_{10}, then we have

$$z_1 = V_1/V_{10} ; \quad z_2 = V_2/V_{10} . \tag{25}$$

It is assumed that z_1 does not change in the course of the reaction, whereas z_2 increases. The conversions of amine and epoxy equivalents, defined as p_A and p_B, are related: $p_A = \zeta p_B$. Then

$$z_1 = \overline{V}_1 \overline{V}_{10} = \overline{M}_{n,1}/\overline{M}_{n,10} = \left(A_4 + B_2\right) / \left(A_4 + B_2 - 4p_A A_4\right) . \tag{26}$$

From the definition of ζ it follows that

$$z_1 = \frac{1 + 2\zeta}{1 + 2\zeta - 4p_A} . \tag{27}$$

For stoichiometry system $\zeta = 1$, $p_A = p_B = \rho$ and $z_1 = 1/(1 - 4\rho/3)$.

The next step consists of determining the parameter χ. It can be found from the cloud points and from the equation for excess chemical potentials. We can express the free energy as

$$\Delta G_V = \left(\frac{RT}{\overline{V}_{10}}\right) \left(\Delta\mu_1\varphi_1 + \Delta\mu_2\varphi_2\right) , \tag{28}$$

where

$$\Delta\mu_1 = \frac{1}{z_1} \ln \varphi_1 + \left(\frac{1}{z_1} - \frac{1}{z_2}\right) \varphi_2 + \chi\varphi_2^2 , \tag{29}$$

$$\Delta\mu_2 = \frac{1}{z_2} \ln \varphi_2 + \left(\frac{1}{z_2} - \frac{1}{z_1}\right) \varphi_1 + \chi\varphi_1^2 . \tag{30}$$

In equilibrium at the cloud point the following relations are valid:

$$\Delta\mu_1^{\alpha} = \Delta\mu_1^{\beta} , \tag{31}$$

$$\Delta\mu_2^{\alpha} = \Delta\mu_2^{\beta} , \tag{32}$$

where superscript α relates to the continuous phase (usually epoxy polymer) and β to the disperse phase (rubber). For low rubber concentration φ_{20} the amount of phase β at the cloud point is small, and $\varphi_2^{\alpha} = \varphi_{20}$, $\varphi_1^{\alpha} = 1 - \varphi_2^{\alpha}$, and $\varphi_1^{\beta} = 1 - \varphi_{20}$.

Equations 29–32 contain two unknown values: φ_2^{β} ($\varphi_1^{\beta} = 1 - \varphi_2^{\beta}$) and χ. By solving these equations one can find the value of χ for any φ_{20}. Now, the parameters found can be put into Eq. 22, from which the binodal and spinodal may be found in the usual way. The authors [77] have computed conversion degree–composition curves.

To describe phase separation the dependence of free energy density on the composition (at any conversion degree) is considered. To derive the driving potential of phase separation, it is necessary to concentrate on the metastable region of the phase diagram. Let φ_{2C} be the true concentration of rubber in the continuous phase at some conversion degree, and φ_2^{α} and φ_2^{β} be the conjugated points of the binodal (equilibrium phase compositions). The sign of being in the metastable region is $\varphi_{2C} > \varphi_2^{\alpha}$. The tangential line at φ_{2C} gives the free energy density of a solution, separated from a matrix of composition φ_{2C}. The

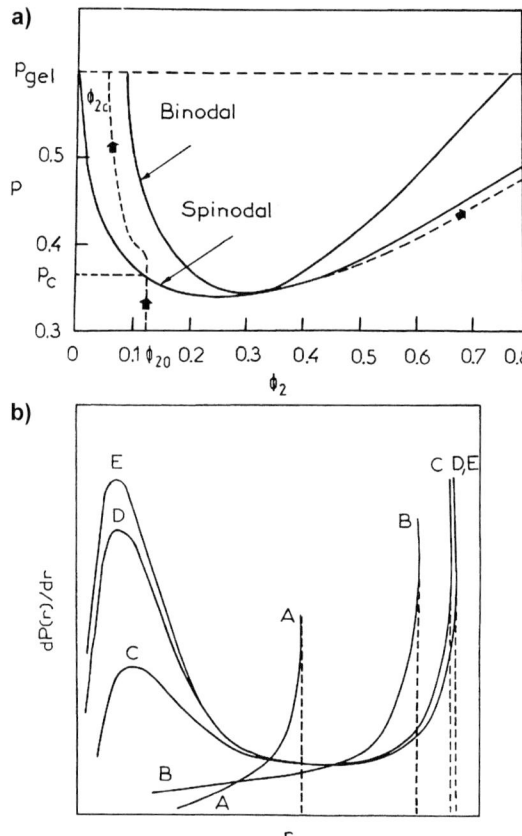

Fig. 7 a Free energy of mixing per unit volume as a function of rubber volume fraction, for a reaction where phase separation is taking place. φ_2^α and φ_2^β are the conjugated points of the bimodal curve, φ_{2C} is the actual rubber concentration in the continuous phase, φ_{2N} represents the actual rubber concentration of the segregated phase, and ΔG_N is the free energy change per unit volume, associated with the phase separation process [77]. **b** Particle size distribution of dispersed domains for different conversions of the thermosetting matrix: curve A, $p = 0.255$; curve B, $p = 0.295$; curve C, $p = 0.395$; curve D, $p = 0.495$; curve E, $p = 0.533$ [77]

vertical distance between the tangential line at φ_{2C} and the free energy curves (Fig. 7b) represents the change in free energy during the creation of a new phase of any composition [77]. For a particular composition, for example, φ_{2N}, the maximum value of free energy ΔG_N is obtained. Thus, the value of φ_{2N} really represents the composition of a segregated phase, whereas ΔG_N corresponds to changes of free energy. As one can see, φ_{2N} is always greater than φ_2^β. The experimental data testify that the systems under consideration are characterized by separation according to the nucleation and growth mechanism. With the exception of incompatible systems, which may separate even at a very small conversion degree, the possibility of coalescence of dispersed domains is restricted by the strong increase in the system viscosity.

During nucleation the change in free energy when a spherical particle with composition φ_{2N} is formed is given by

$$\Delta G = (4/3)\,\pi r^3 \Delta G_N + 4\pi r^2 \gamma ,\tag{33}$$

where r is the radius of a particle (domain) and γ is the interfacial tension. The maximum value of ΔG_{crit} is reached at the critical domain radius r_{crit}. From Eq. 33 we have:

$$r_{crit} = 2\gamma\,|\Delta G_N| ,\tag{34}$$

$$\Delta G_{crit} = \frac{16\pi\gamma^3}{3\,|\Delta G_N^2|} .\tag{35}$$

The growth of dispersed domains with $r = r_{crit}$ diminishes free energy, as is shown in Fig. 8.

The rate of homogeneous nucleation from condensed phases is written as:

$$\frac{dP(r_{crit})}{dt} = N_0 D \exp\left(\frac{\Delta G_{crit}}{k_B T}\right) ,\tag{36}$$

where $P(r_{crit})$ is the volumetric concentration of particles with critical radius r_{crit}, D is the diffusion coefficient for rubber in an epoxy network, N_0 is the pre-exponential factor, and k_B is the Boltzmann constant.

To calculate the nucleation rate, one has to evaluate γ and D (N_0 is found from experimental data on particle concentration). The value of γ was evaluated [77] as the difference between the surface tension of epoxy resin and that of rubber, i.e., by modeling the epoxy-enriched phase as a pure epoxy resin and the rubber-enriched phase as a pure rubber. Such simplification has no

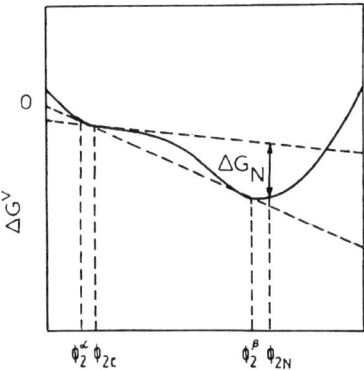

Fig. 8 Different contributions to the free energy change associated with the formation of spherical domains [77]

effect on the final results. The diffusion coefficient is determined using the Stocks–Einstein equation:

$$D = \frac{D_0 T}{\eta (p, T)} , \tag{37}$$

where η is the viscosity of the continuous phase, which depends on both temperature and conversion degree. At $p = p$ (gel), $\eta(p, T) \to \infty$, $D \to 0$, and nucleation is stopped. Really, near the gel point the strong increase in viscosity leads to very small nucleation rates.

When the system passes through a metastable region, the particle growth onsets, whose magnitude is determined by the value of the driving force $(\varphi_{2C} - \varphi_2^\beta)$. This force should transfer the system to the equilibrium state. Let $dP(r')$ correspond to the change in the number of particles per unit volume the when conversion degree changes from p' to $p' + \Delta p$. At $p = p'$ we have a set of particles with radius $r' = r_{crit}(p')$. For $p > p'$ this radius will growth up to some value $r' > r_{crit}(p')$. The value of $dP(r')$ does not depend on time. The growth rate is defined as an increase in the volume fraction of particles per unit time and should be proportional to the interfacial area per unit volume and to the driving force:

$$(4\pi/3) \, dP \left(r'\right) \left(dr'\right)^3 / dt = K_v 4\pi \left(r'\right)^2 dP \left(r'\right) \left(\varphi_{2C} - \varphi_2^\alpha\right) , \tag{38}$$

where K_v is the mass transfer coefficient for the sphere in a stationary medium $(K_v = D/r')$. When $D \to 0$ for $p \to p$ (gel), the rate of growth is restricted by gelation.

Based on this, the structure of the system can also be described in the following way. The volume fraction of evolved dispersed phase, $dV_D(p)$, in the region of conversion degrees from p to $p + \Delta p$, includes new particles which appear at P^* plus an additional fraction of particles, which are formed at $p < p'$:

$$dV_D \left(p\right) = \left(4\pi/3\right) \left[r_{crit} \left(p\right)\right] dP \left(r_{crit}\right) + 4\pi \int_{p'<p} \left(r'\right)^2 dP \left(r'\right) dr' . \tag{39}$$

All the material segregated in the region of conversion degrees from p to $p + \Delta p$ has the composition φ_{2N}. The total concentration of particles of dispersed phase per unit volume at any conversion degree is given by

$$P = \int_{p' \leq p} dP \left(r'\right) . \tag{40}$$

In this case, the function describing the distribution of particle size in the dispersed phase may be written as $dP(r)/dr$. The volume fraction of dispersed phase at any conversion P is

$$V_D = (4\pi/3) \int_{p' \le p} (r')^3 \, dP(r') . \tag{41}$$

A volumetric average diameter of dispersed phase particles is calculated as

$$\overline{D} = (6V_D/\pi P)^{1/3} . \tag{42}$$

The average rubber concentration in the dispersed phase is

$$\bar{\varphi}_{2D} = (1/V_D) \int_{p' \le p} \varphi_{2N}(p') \, dV_D(p') . \tag{43}$$

If φ_{20} is the initial volume fraction of rubber in the system, then its concentration in the continuous phase may be found from the equation

$$\varphi_{2C} = \frac{\varphi_{20} - V_D \bar{\varphi}_{2D}}{1 - V_D} . \tag{44}$$

The applicability of the model discussed above to the processes of phase separation was proved for the thermocuring system consisting of epoxy resin based on the bisphenol A diglycidyl ester of diaminodiphenyl sulfone, in which the statistical copolymer was dissolved. For this system, the binodal and spinodal have been constructed and the dependencies of nucleation and growth rates were calculated (Fig. 7b). The amount of rubber in the continuous phase φ_{2C} does not change significantly up to the conversion degree $p = 0.5$ due to a rapid increase in viscosity in this region. When the compatibility of the resin and the rubber diminishes (as in the case of the same rubber with a smaller amount of acrylonitrile), the phase separation proceeds in the medium with lower viscosity and the trajectory of φ_{2C} in the metastable region fast approaches binodal due to the high diffusion coefficient. A small driving force $(\varphi_{2C} - \varphi_2^\alpha)$ results in a small rate of growth whereas the nucleation rate remains high. As a result, the bimodal distribution of particle size onsets. At the early stage of phase separation a very sharp distribution of particles is observed (curve A). In the course of the reaction the particles grow fast, with small particles appearing in the front tail of the distribution (curve B). As the gel point approaches, the growth of particles of large size slows down and the concentration of small particles increases (curves C–E). The reason for this is the fact that the growth rate is inversely proportional to the dimension of dispersed domains. Generally, it is established that the decreasing compatibility, i.e., increasing parameter $\chi < 0$, leads to (1) increasing concentration of the dispersed phase, (2) diminished amount of rubber in the continuous phase, and (3) diminished average concentration of small particles in the total distribution. All these effects result from the process far from the gel point in low-viscosity medium.

The analysis of the interfacial tension effect on the distribution has shown that change from 2γ to $\gamma/2$ had no significant effect on the final distribution. This is explained by the fact that in the metastable region, the $\Delta G_{crit}/RT$ factor (Eq. 36) is very small and has no influence on the nucleation rate. On the other hand, although r_{crit} is directly proportional to γ (Eq. 34), the mass transfer coefficient is inversely proportional to the particle size. As a result, the final size is almost independent of the critical radius r_{crit}.

Rosenberg et al. [78–80] have thoroughly investigated the processes of microphase separation induced by the reaction of curing polymeric–oligomeric systems. The processes of structure formation in these systems proceed according to the mechanism of nucleation and growth. The main attention was paid to the role of the chemical kinetics of reaction and the physical kinetics of phase separation in the formation of the cured system. The thermodynamic analysis [81] takes into account that the process of curing is not at equilibrium and the nonuniformity in the component distribution. It was shown that the phase diagram of curing a two-component system may be well described by the Flory–Huggins theory, being constructed in coordinates of conversion degree at the onset of phase separation (cloud-point conversion) and composition. However, in the real case the forming system is polydisperse and for thermodynamic analysis it is necessary to account for the molecular mass distribution, differences in the thermodynamic interaction parameters χ_{ij}, and volume changes in the course of the reaction. For curing systems, the author proposes to use the generalized Flory–Huggins equation in the following form:

$$\frac{\Delta g_m}{RT} = \sum_i \frac{1}{V_i} \varphi_i \ln \varphi_i + \frac{1}{V_r} \sum_{i \neq j} \chi_{ij} \varphi_i \varphi_j , \tag{45}$$

where φ_i and φ_j are volume fractions of components, and V_i and V_r are the molar volumes of the i-th and smallest components. This equation includes a set of interaction parameters, which may be calculated from the solubility parameters. The construction of the phase diagram of the curing system by using this equation for a multicomponent system is a nontrivial task. To solve the problem the author introduces the concept of the free energy of phase separation by the formation of the stable two-phase system, Δg_{ps} from one-phase solution. This value may be calculated from the free energies of mixing of heterophase G_m^2 and homophase G_m^1 systems that are formed by mixing n components and reduced to the volume unit.

If the phase separation is possible, the free energy of phase separation is negative and characterized by a deep minimum. Because of this the unknown variables of the equation, which are necessary for constructing phase diagrams, may be found by minimization of the function Δg_{ps}. This term, as distinct from the free energy of mixing, allows in the closed form the condition of the phase equilibrium (bimodal equation) to be recorded for multicomponent polymer systems. The bimodal is found by a search of the global minimum of Δg_{ps}.

As Rosenberg emphasizes the application of the approach developed is based on the assumption that phase separation proceeds at equilibrium conditions. For curing systems this condition is unrealistic because of the simultaneous proceeding of both the chemical and physical processes. The criterion of the equilibrium of the process is determined by the ratio of the rate of the chemical reaction rate and mutual diffusion coefficient at which the process may be considered as equilibrium. In his work Rosenberg also proposes models which describe the nucleation and growth of particles of the dispersed phase. These models account for the effects of the reaction rate and nonuniform space distribution of components in the course of phase separation. The models allow the final morphology of the system to be described.

2.3.2
Spinodal Decomposition

The possibility of another mechanism of phase separation in IPNs, spinodal decomposition, was discovered in 1984–1985 [5, 53, 82–88]. Studying various semi- and full IPNs we have discovered that the microheterogeneous structure of IPNs arises as a result of microphase separation proceeding according to the spinodal mechanism of phase separation. Spinodal decomposition of phase separation arises in the region of the unstable state on the phase diagram. The main problem consists of the study of microphase separation proceeding simultaneously with the chemical cross-linking reaction. In this case curing occurs simultaneously with the growth of conversion degree of the components and growth of the molecular mass of the chain fragments, which become incompatible at a certain conversion degree.

It was already mentioned that the thermodynamic stability of the multicomponent systems is determined by the concentration dependence of the Gibbs free energy. If $\partial^2 G/\partial\varphi^2 > 0$ (as in the region between spinodal and binodal), the system is unstable only to large concentration fluctuations. The system represents the supersaturated solution, which decomposes with the formation of nuclei, rich in the dissolved component, surrounded by the dispersion medium. If $\partial^2 G/\partial\varphi^2 < 0$ (inside the spinodal region), the system is unstable to small fluctuations, and the supersaturated solution begins to separate into two phases simultaneously in the whole volume of the system without forming the nuclei. Spinodal decomposition is the kinetic process of spontaneous formation and continuous growth of one phase in the instable "mother" phase. This process is determined by the appearance of fluctuations of small amplitude, which leads to statistically continuous growth of the second phase, characterized by sinusoidal changes of the composition. The decomposing system is characterized by the high degree of interpenetration of evolving phases. If during nucleation the diffusion flows are diminishing the composition fluctuations, during spinodal decomposition the directions of diffusion flows reverse. The intensity of composition fluctuations increases, and the instability of the system also increases. The spinodal decomposition occurs inside the spinodal region of the phase diagram. In this region phase separation leads to the formation of microregions with compositions which deviate from the system composition by a very small amount. The thermodynamic interaction parameter and the thickness of the transition layers are important quantities to enter in theoretical equations. In spite of the fact that in the literature there were many theoretical and experimental works on spinodal decomposition in polymer blends, the first work on decomposition in IPNs appeared only in 1984 [84].

The difficulty of studying this process is connected to the simultaneously proceeding chemical reaction. Because of this, for our investigations we have chosen systems where the chemical reaction proceeds slowly in order to allow the maximum time for phase separation. Consider the process involving a semi-IPN based on cross-linked PS and PBMA, which show limited compatibility at low concentrations of components at low MM, but are totally incompatible at higher MM of components [88]. The transition from compatibility to incompatibility proceeds in the course of this reaction as the MM grows. The experimental data show that the incompatibility appears at the very early stages of reaction (for conversion values of 1–8%). Increase in the fraction of PBMA in the system leads to lowering of the conversion degree, at which phase separation begins. Low miscibility of both components follows from their phase diagrams. When analyzing the results from investigation of phase separation one has to keep in mind the great complexity of the system under study. The process proceeds in a system where continuous changes of composition take place. As a result, the thermodynamic compatibility also changes continuously. How can we prove which mechanism controls the phase separation in the system under study? For both mechanisms, the intensity of light scattering (the main method of investigation), I, changes with time. For nucleation and growth the following relation is valid:

$$I \propto t^2 . \tag{46}$$

For spinodal decomposition another relation works:

$$I \propto e^t \tag{47}$$

or

$$\ln I\left(\beta, t\right) = \ln I\left(\beta, 0\right) + 2R\left(\beta\right) t , \tag{48}$$

where $R(\beta)$ is the amplification factor, which is related to the wave number β:

$$R\left(\beta\right) = M\left(\frac{\partial^2 f}{\partial \varphi^2}\right)\beta^2 - 2M\beta^4 = D_{\text{app}}\beta^2 - 2M\beta^4 , \tag{49}$$

where D_{app} is the apparent diffusion coefficient, $D_{app} = D_c(T)\left(\frac{\chi - \chi_s}{\chi_s}\right)$, D_c is the translation diffusion constant, χ_s is the interaction parameter at spinodal temperature; M is the diffusion mobility and

$$\beta = \frac{4\pi}{\lambda}\sin\left(\frac{\theta}{2}\right). \tag{50}$$

For high molecular mass systems with restricted mobility, the region of the metastable state (between binodal and spinodal) is very wide. To reach the region inside spinodal and to initiate the spinodal decomposition, it is necessary to rapidly force the system through the metastable region. Next, Figs. 9 and 10 show the time dependence of the intensity of scattered light ($\ln I = f(t)$) for the semi-IPN (styrene–DVB copolymer/PBMA with 10% by mass) undergoing phase separation for incident angles 10° (curve 1) and 20° (curve 2) at 333 K (a), 343 K (b), and 363 K (c) [88]. As one can see, the time dependence of $\ln I$ is linear and it remains at elevated temperatures even for the systems with higher PBMA content, where the duration of phase separation does not last longer than 200–300 s. At lower temperatures and for lower content of PBMA (with phase separation duration of 700–2000 s), this dependence becomes nonlinear. However, in these curves two branches can be distinguished. The first branch corresponds to the initial stage of phase separation. Then, there is an inflection point, after which the second, more prolonged branch, can be seen.

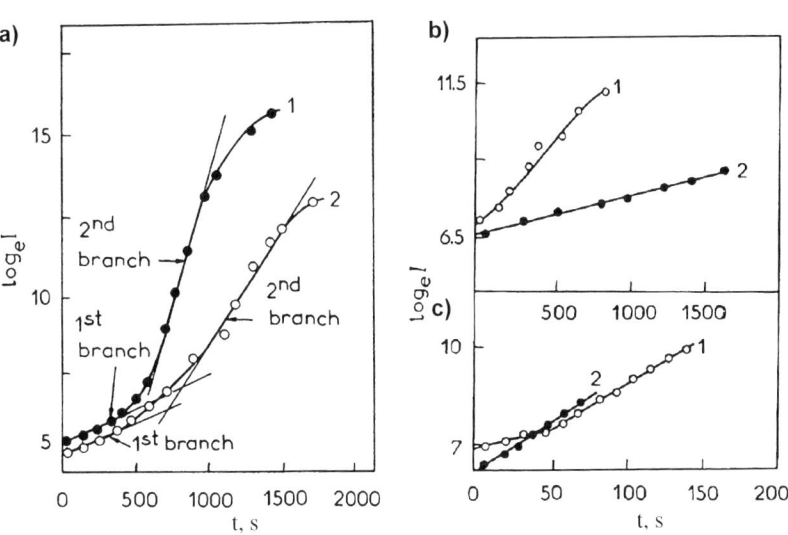

Fig. 9 Dependence of the logarithm of the scattered light intensity (I, in arbitrary units) on time (t) for the styrene–DVB–PBMA IPN with 10 mass % of PBMA undergoing phase separation for incident angles 10° (curve 1) and 20° (curve 2) at **a** 333 K, **b** 343 K, **c** 363 K [88]

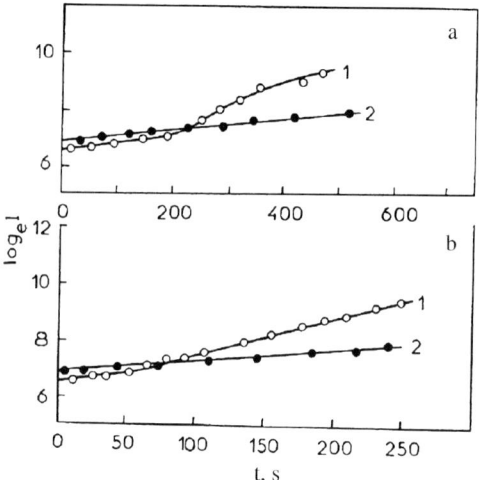

Fig. 10 Dependence of the logarithm of the scattered light intensity on time for the styrene–PBMA IPN with 20% of PBMA undergoing phase separation for incident angles 10° (curve 1) and 20° (curve 2) at **a** 333 K **b** 343 K [88]

Cahn's theory was developed for the initial stages of phase separation (remember that the final result of the equilibrium phase separation does not depend on the mechanism and is determined only by the phase diagram of the system).

We can apply this theory to our case provided phase separation takes no more than 200–300 s (the first linear branch). In our case, during the reaction time when phase separation was observed, the composition of the system changes only slightly and can be considered as constant. This fact gives us the right to apply Cahn's theory for describing the phase separation in our system. However, we believe that Cahn's theory can also be formally applied to the second linear branches in Fig. 9.

Figure 11 shows the dependence of $R(\beta)/\beta^2$ on β^2 for both branches of the system with 10 mass % PBMA. It can be seen that the linearity of these dependencies is improved as the temperature increases. If that dependence can be approximated by a straight line, some important theoretical parameters can be calculated. From the dependence of the slope of the $\ln I$ vs t plot, the amplification factor can be calculated, and from the dependence of $R(\beta)/\beta^2$ on β^2 the interdiffusion coefficient can be found, as well as the most important characteristics—the optimum wavelength of spinodal decomposition λ_m:

$$\lambda_m = 2\pi/\beta_m . \tag{51}$$

Thus, the value of λ_m characterizes the size of the microheterogeneity regions. The value of λ_m is calculated from the dependence mentioned above. The interdiffusion coefficients and the sizes of the microheterogeneity regions for the systems with 10, 15, and 20 mass % of PBMA are given in Table 1 [88].

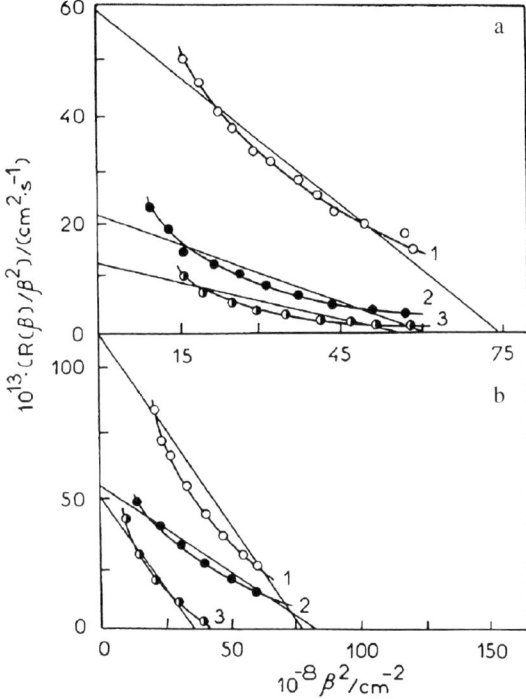

Fig. 11 Dependence of $R(\beta)$ on β^2 according to Eq. 49 for the semi-IPN with 10 mass % of PBMA as calculated for the first (**a**) and second (**b**) linear branches in Fig. 9 at 363 K (1), 343 K (2), and 333 K (3) [88]

One can see that the size of the microheterogeneity regions is nearly the same for all temperatures. The calculations of the interdiffusion coefficient have shown that it is negative, which corresponds to the spinodal mechanism. With growing temperature D increases, i.e., interdiffusion becomes easier due to the lower viscosity of the system. The absolute values of D for the second stage are higher. This means that the effective diffusion is different for various stages of the decomposition. In the second stage the diffusion proceeds faster. The activation energies of diffusion for the second stage (calculated from the temperature dependence of the interdiffusion coefficient) for the system with 10 mass % PBMA are two times less than the activation energy for the first stage ($E_A = 26.96$ kJ mol^{-1}). At the same time, for the system with 20 mass % of PBMA the activation energy is nearly the same for both stages ($E_A = 16.62$ kJ mol^{-1}). This fact proves the equivalence of the effective diffusion for both stages.

The data on the microphase structure of the semi-IPN show that the phase separation proceeds according to the spinodal mechanism in spite of the simultaneously proceeding chemical reaction. Thus, the process is a nonequilibrium one. The peculiarity of this process is in its two-stage nature. It

Table 1 Interdiffusion coefficients D and microheterogeneity region sizes (λ_m) as functions of temperature and composition for semi-IPNs (PS–PU) with different contents of PBMA [88]

Temperature, K	$D \cdot 10^{13}$ (cm$^2 \cdot$s^{-1})		λ_m (μm)	
	Stage 1	Stage 2	Stage 1	Stage 2
		10 mass % PBMA		
333	13	50	1.2	1.1
343	22.5	56	1.2	1.0
363	61.5	113	1.0	1.4
		20 mass % PBMA		
333	35	46	1.6	1.4
343	42	54	1.5	1.1

may be assumed that the initially "pure spinodal mechanism" becomes more complicated in time due to the simultaneously proceeding process of phase separation according to the nucleation mechanism. However, this process does not lead to full separation into two independent phases due to chemical cross-linking and network formation. As a result, the system with diffuse microregions is formed. It is worth noting that the superposition of chemical reaction and phase separation in many cases makes it difficult to determine the mechanism of phase separation; in some cases it is possible for one mechanism to be replaced by another. As a result, thermodynamically unstable diffusion microregions of incomplete phase separation are created, so that the segregation degree in such systems is not an equilibrium value.

As will be shown later, the phase separation begins at very low degrees of conversion and is enhanced by increasing MM and volume fraction of copolymer. The kinetics of IPN formation determines the onset of phase separation and influences strongly this process as a whole, whereas phase separation does not influence the kinetic curves. This fact may serve as an additional confirmation of the assumption that the phase separation proceeds according to a spinodal mechanism, because in this case the compositions of two evolving phases are very close.

The phase separation during synthesis of poly(ether imide)/epoxy semi-IPNs was studied [89] using the morphological data for various compositions of the systems and turbidity experiments (determination of cloud points). The phase separation behavior during curing was followed by observing the morphology changes and the light scattering. Experimental data suggest that the phase separation proceeds in two stages, leading to the formation of so-called dual-phase morphology. The authors observed spinodal structures and believe that the process is governed by the spinodal mechanism, although the typical experiments allowing us to establish the mechanism have not been performed. They conclude that because of the spinodal mechanism and the low viscosity of the system, macro-scale phase separation of domains of various compositions will occur. As the reaction proceeds, there occurs an abrupt change in the equilibrium composition of phases in a very short time. This abrupt change is similar to the effect of the two-step temperature quenching in polymers with UCST behavior. This in turn results in the dual-phase morphology. The first phase-separated composition has higher viscosity and demonstrates the jump on the UCST curve, which induces a secondary phase separation within both the domain and the matrix. We see, however, that no attempt has been made to describe the spinodal decomposition within the framework of current theories and to determine the typical values characterizing this process, as was done in earlier works.

For simultaneous semi-IPNs made from PU and PS, the kinetics of phase separation was studied using optical microscopy completed by image analysis [90]. The development of a nodular structure was observed. A thermodynamic approach has allowed us to establish that the diameter of the phase-separated species was the result of the competition between the kinetics of network formation and the kinetics of phase separation. The mechanism of phase separation was not discussed.

In connection with the above discussion, an important question arises concerning the thermodynamic state of systems with various degrees of phase separation (segregation degree) [90]. Numerous experimental data show that segregation during microphase separation depends on the cross-linking degree, the reaction kinetics, the composition, the entanglements between growing chains before phase separation, etc. It is evident that after the transition of the system from the one-phase state into the metastable and unstable regions, the stable chain entanglements (physical or chemical) do not allow the full separation of network fragments and the system stays in the state of "forced" miscibility (compatibility). This state makes an additional contribution to the free energy of the system.

It is evident that the contribution of the elastic energy in multicomponent polymer systems is small at the onset of phase separation. It increases with growth in the composition difference of phase-separated microregions. As a result, the system is stabilized when the thermodynamic force of phase separation is balanced by the elastic forces from entanglements of the network fragment. The system remains in a state of incomplete phase separation. The stability limit of such a polymer system will be characterized by the curve below the chemical spinodal. For the system with UCST the real spinodal is located below the chemical spinodal. It is important to note that for some systems, including IPNs, the real spinodal is so removed from the chemical one that the unstable state cannot be reached.

Analysis of the behavior of some IPNs shows that such a system may not only be in a state of "forced" compatibility, but also in a state of "forced" microphase separation [86]. If such systems have an UCST, the end of the transition to the one-phase system corresponds to the curve located above the chemical spinodal (upper real spinodal). Its position is determined by the values of the elastic forces which are needed for the system to transform from a state with microphase separation to the one-phase state. Such a picture was observed experimentally [91]. In this way, the transition of the system from a one-phase to a two-phase state is hindered at a particular stage of IPN formation (lower chemical spinodal), whereas the transition from the two-phase state to the one-phase state is hindered as well (upper real spinodal). Thus, the state of the system is determined by its synthesis. This synthesis means that IPNs have some regions near the chemical spinodal where a state with incomplete phase separation may exist.

The spinodal as a curve, which restricts the unstable region, becomes "diffuse" over a wide temperature range. In this range the states, which can be related neither to one-phase nor to two-phase systems, appear. This question was qualitatively discussed in detail from the point of view of equilibrium and nonequilibrium fluctuations arising in the system [90]. The phase separation is incomplete due to kinetic stability in the region between the chemical and the real spinodals. There are no analogues to such a system in the traditional concept of one- and two-phase systems.

The effect of cross-linking density on phase separation was studied for semi-IPNs made from deuterated PS (PSD) and poly(vinyl methyl ether) (PVME) by polymerizing a styrene–DVB mixture that contains dissolved PVME, and those made from PS and PSD [92]. In the latter case it was possible simultaneously to estimate both the cross-linking degree effect and the chain length of the linear polymer. To measure the concentration fluctuations in the systems, small-angle neutron scattering (SANS) was used. The initial linear polymers were miscible and because of this, the effect of cross-linking on phase separation could be seen very clearly. The interaction parameter χ was changed by increasing the temperature during the measurements of zero-angle scattering and correlation length. It was found that when the temperature increases, the scattering intensity increases as well, indicating that the spinodal temperature is being approached. Samples with the lowest cross-linking density have scattering similar to that of a compatible polymer blend. As the temperature is increased, the scattering also increases, showing LCST behavior. The systems with the highest cross-linking degree remain a single phase upon polymerization. During polymerization, the cross-links seem to push the system toward phase separation.

Theoretically, the cross-linking degree is taken into consideration in the following way, based on the Flory–Rehner theory [92]. The coexistence curve may be calculated from the chemical potentials of the linear chain inside and outside of the network. The chemical potential of the linear chain is given by:

$$\mu_b = \frac{A\varphi_s^{2/3}\varphi^{2/3}}{v_a N_c} - \frac{B\varphi}{v_a N_c} + \frac{\varphi}{v_b N_b} + \frac{\ln(1-\varphi)}{n_b N_b} + \frac{\chi\varphi^2}{v_0}. \tag{52}$$

The spinodal condition $\partial^2(\Delta F/kT)/\partial\varphi_2 = 0$ is described by the following equation:

$$\frac{\partial^2 \left(\Delta F/kT\right)}{\partial\varphi^2} = \frac{B}{v_a\, N_c\varphi} + \frac{A\varphi_s^{2/3}}{v_a N_c\varphi^{5/3}} + \frac{1}{v_b N_b\left(1-\varphi\right)} - \frac{2\chi}{v_0} = \frac{k_n}{S(0)}. \tag{53}$$

In these equations v_a, v_b, and v_0 are the molar volumes of the network repeating unit, N_c is the number of repeating units in the network between cross-links, χ is the Flory–Huggins parameter, φ is the volume fraction of the network, φ_s is the network volume fraction in relaxed state, usually taken to be the value at which the network was formed and N_b is the number of repeat units in the linear chain. Constants A and B follow from the rubber elasticity theory, usually $A = 1$ and $B = 2/f_c$, where f_c is the functionality of the cross-links, k_n is the constant that determines the amount of contrast between the two components and the radiation type, and $S(0)$ is the zero-angle scattering intensity.

These equations allow one to calculate the phase diagrams (Fig. 12). Equation 53 predicts that cross-links should inhibit concentration fluctuations. The experimental data on scattering intensity were compared with theoretical predictions.

Generally, the level of investigation of the spinodal decomposition in IPNs does not correspond to the present state of the theories, developed for describing spinodal decomposition in blends of linear polymers [52]. Unfortunately, up to now there are no quantitative theories of spinodal decomposition in IPNs. The only work was done by Binder and Frisch [93].

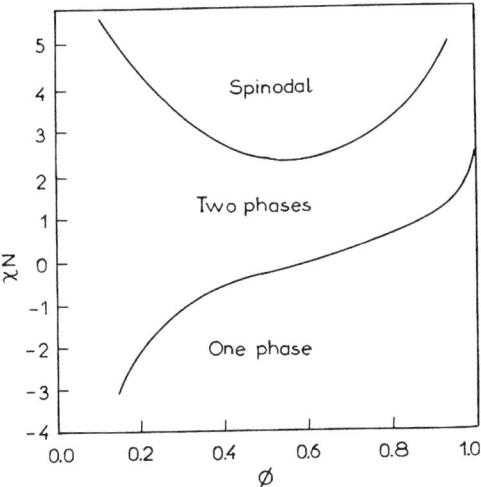

Fig. 12 Cloud point (*lower line*) and spinodal (*upper line*) curves for a polymer network (concentration φ) polymerized in the presence of a linear polymer B (concentration $(1-\varphi)$), $N_c = N_b = N$, $\Phi s = \Phi$ [92]

2.3.3
Binder–Frisch Approach

A phenomenological theory for chemically quenched binary IPNs of both simultaneous and sequential type and for semi-IPNs was formulated by Binder and Frisch [93]. Phase separation proceeding during reaction is hindered by simultaneous cross-linking termed "chemical quenching". The authors analyze the case of deep quenching when the reaction rate essentially exceeds the rate of relaxation of the system into the equilibrium state. The free energy functional for these systems strongly depends on the method of IPN formation (sequential or simultaneous). Starting from the Flory theory of rubber elasticity and the Flory–Huggins interaction parameter χ, the case of weakly cross-linked networks was considered, in which the effective chain length between cross-links is very large. Individual networks are assumed to be described by the mean field statistical theory of rubber elasticity. The authors treat as chemically quenched IPNs those systems whose characteristic times, θ_A and θ_B for curing the networks A and B, are short compared to the characteristic time of growth of the initial phase instability τ_{AB}. The theory for the case when $\theta_I > \tau_{AB}$ is only possible to construct within the framework of the time-dependent, molecular statistical mechanical theory.

To find the expression for the free energy functional, the authors assume that the free energy of IPNs consists of several parts: (1) elastic free energy and entropy connected to the cross-links of two networks A and B, which can be found from several theories; (2) interaction energy between two kinds of monomers; (3) entropy of mixing if one of the two species consists of linear polymer (semi-IPN); and (4) free energy of mutual entanglement between two networks. Analyzing the corresponding expression for each part of the free energy, the authors have derived expressions for all four values. In the process they have taken account of the values characterizing the deformation ratios with respect to a reference state, chain dimensions, the number of elastically effective chains $\nu(\varphi_0)$, where φ_0 is the volume fraction of the monomers of species A in the mixed system, the number of sites in the Flory–Huggins lattice, etc.

On the basis of the equations obtained, Binder and Frisch found the expressions for spinodal curves and critical points. For simultaneous IPNs the equation for the stability limit of the homogeneous phase (i.e., for spinodal curve, $\chi = \chi_S$) was found to be:

$$0 = \left(\frac{\partial \mu}{\partial \varphi} \right)_{\varphi_0} \tag{54}$$

$$= \frac{\varphi_{A^*}^{2/3}(\varphi_0)}{3N_{AC}(\varphi_0)} \varphi_0^{-5/3} + \frac{B_A}{N_{AC}(\varphi_0)} \varphi_0^1 - \frac{\left(1 - \varphi_{B^*}\right)^{2/3}}{e} N_{BC} + \frac{B_B}{N_{BC}} \left(1 - \varphi_0\right)^{-1},$$

where φ_A and φ_B are a function of φ_0, N_{AC} and N_{AB} are the functions connected with the number of sites in the lattice for A and B species, $B = 2/f$, where f is the functionality of cross-links, and φ^* denotes hypothetical volume fractions that have to be chosen after the cross-linking to eliminate the elastic force of cross-links. This equation allows one to construct the spinodal curve $\chi = \chi_S(\varphi_0)$, in terms of the functions $N_{AC}(\varphi_0)$, $\phi_{A^*}(\varphi_0)$, $\varphi_{B^*}(1 - \varphi_0)$, and $N_{BC}(1 - \varphi_0)$.

For various elasticity theories, the qualitative "phase diagrams" have been constructed as functions of the Flory–Huggins interaction parameter (Fig. 13) [93]. Case a describes simultaneously cross-linked IPNs from the point of view of James–Guth theory, case b from the point of view of Flory theory, while case c relates to sequentially cross-linked IPNs using Flory theory.

Broken and dash-dotted curves show two possible cases for $\chi(\varphi)$. Each case is realized depending on the function $N_{AC}(\varphi)$. In all the cases only the stability limits of the homogeneous phases are shown and no attempt has been made to construct the coexistence curves, where the free energies of the homogeneous phase and the spatially modulated phase (or unmixed phase) are equal.

The dynamic response and the spinodal decomposition of IPNs have also been considered (dynamics of the volume fraction fluctuation in the networks) based on the application of the equation for the relaxation of volume fraction fluctuations. According to the theory, in the long-wave limit, the fol-

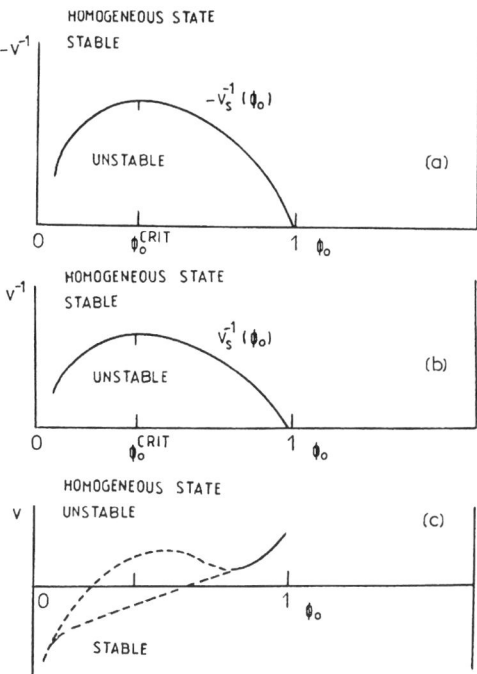

Fig. 13 Qualitative phase diagrams of the IPNs shown as a function of χ [93]

lowing expression is valid:

$$\sigma_A^2 q^2 \gg N_{AC}^{-1}, \quad \sigma_B^2 q^2 \ll N_{BC}^{-1}, \tag{55}$$

where σ_A and σ_B are the dimensions of statistical segments in chains A and B, and q is the wave vector characterizing the rate of fluctuation growth. It was found that the composition fluctuations begin to grow when the values of wave vectors are less than q_c:

$$q_c = \sigma^{-1} \left[-\left(\frac{\partial^2 f}{\partial \varphi_2^2}\right)_{\varphi_0} \frac{\varphi_0 (1 - \varphi_0)}{K(\varphi)} \right]^{1/2}, \tag{56}$$

where φ is the volume fraction of one of the networks, φ_0 is the fraction of monomers of network A, and K is a coefficient depending on φ_0. It is assumed that $\sigma_A = \sigma_B = \sigma$. In this case the phase separation stops at the initial stages by forming the modulated structures whose spatial period is much larger than the dimension of the network cell.

For $q \leq q_c$, the fluctuations grow and their maximum growth occurs at

$$q_m = \frac{1}{\sqrt{2}} q_c . \tag{57}$$

The fluctuation growth rate is

$$\left| \tau_{q_m}^{-1} \right| = \frac{\Lambda \varphi_0 (1 - \varphi_0)}{4\sigma^2 K(\varphi_0)} \left[\left(\frac{\partial^2 f}{\partial \varphi^2}\right)_{\varphi_0} \right]^2, \tag{58}$$

where $f(\varphi_0)$ is the free energy density.

For deep quenching the following expression is valid:

$$q_c = \left[\frac{6\chi^{1/2}}{\sigma_A^2/\varphi_0 + \sigma_B^2/(1 - \varphi_0)} \right]^{1/2}. \tag{59}$$

In this case the local degree of separation of phases may be very high, whereas the distance between microregions is comparable with the dimensions of the network cell (distance between cross-links).

The authors themselves emphasize the two main limitations of this theory. The first limitation concerns the assumption of "chemical quenching", i.e., that the polymerization kinetics and cross-linking are so rapid that any phase separation during the time needed for these processes is negligible. Their estimate of a typical time constant for the development of instability in the linear regime of spinodal decomposition, 10^3 s^{-1}, shows that this limit may not be applicable to all cases of experimental interest. We cannot help accepting this statement. The second limitation is that their treatment does not allow us to describe the periodically modulated structures.

Generally, the following conclusion can be drawn from the analysis of the mechanisms of phase separation in IPNs. The structures arising at different stages of spinodal decomposition are different, and the final structure of the IPN will be determined by the time at which the system loses its mobility due to cross-linking and becomes "frozen" or "chemically quenched". The reaction kinetics (Sect. 5), the composition, and the diffusion are the most essential parameters, which control IPN structure and phase separation. Because these processes are very complicated and interconnected, it is logical to assume that in such systems mechanisms of nucleation and growth or spinodal decomposition can hardly occur in their pure form.

The effects of forced compatibility in IPNs are of great importance for understanding the phase state. Forced compatibility should lead to an additional contribution to the free energy of the system. The contribution of elastic energy is small at the onset of phase separation and increases with the increase in composition difference between phase-separated regions. As a result, the system is stabilized when the thermodynamic driving force for phase separation is balanced by elastic forces from entanglements of network fragments. The system stays in a state of incomplete phase separation. The stability limit of such a polymer system is characterized by a curve below the chemical spinodal. For the systems with UCST, the real spinodal is located below the chemical spinodal. For some IPNs, the real spinodal is so remote from the chemical spinodal that the unstable states cannot be reached.

The analysis of the thermodynamic behavior of some IPNs shows [88, 90] that two states may exist: a state of forced compatibility and a state of forced microphase separation. It should also be noted that the most probable mechanism of phase separation is nucleation and growth for sequential IPNs and spinodal decomposition for simultaneous formation.

2.3.4
Interfacial Region in IPNs

The incomplete phase separation or phase separation due to the spinodal mechanism leads to the formation of a transition zone or an interphase between two evolved phases. The first attempt to estimate the volume fraction of this zone was done in [65, 94].

Nuclear magnetic resonance (NMR) spectroscopy and inverse gas chromatography have been used to study the intermediate glass transition and the activation energy of T_g. This was done using a sequential IPN based on PU as polymer I and PS as polymer II. NMR signals were measured as a function of temperature. The spectrum shape changes from a simple one-component line to a more complicated line as the temperature is increased, indicating the existence of regions of various mobility. Figure 14 shows the temperature dependence of the second moment ΔH_2^2 for IPNs [94]. Three glass transition temperatures can be seen. The transitions near 203 and 333 K coincide with transition points of the individual networks (although this is not typical). However, the appearance of a third, intermediate transition in the range

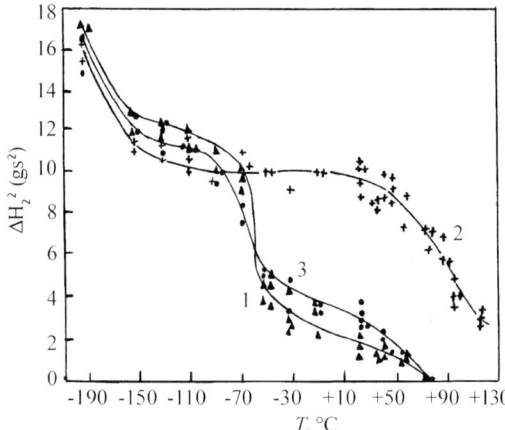

Fig. 14 Temperature dependence of ΔH_2^2 for PU (1); styrene–DVB copolymer (2); IPN at $w_1/w_2 = 0.174$ (3) [94]

288–333 K indicates the presence of an interphase region. The same results were obtained for IPNs based on epoxy resin and allyl derivatives [95].

It is known that for two-component systems, additivity of lines takes place in NMR spectra. Assuming that the same additivity should be observed for the second moment, ΔH_2^2 may be presented in the following way:

$$\left[\Delta H_2^2\right]_{1,2} = W_1 \left[\Delta H_2^2\right]_1 = W_2 \left[\Delta H_2^2\right]_2 = \left[\Delta H_2^2\right] , \tag{60}$$

where W is the weight fraction of the one component and $[\Delta H_2^2]_i$ is a term responsible for interaction, if any. When interaction is absent on the molecular level, this term is equal to zero, a fact which was experimentally confirmed.

Using gas chromatography the excess enthalpy of mixing at the point of equilibrium adsorption may be found from Eq. 2. In this case the excess enthalpy of mixing of sorbate with i-th network and excess enthalpy of mixing of two networks were determined by:

$$\Delta H = R \frac{\partial \left[-\ln V_G P_1^0 - \frac{P_1^0}{RT}\left(B_{11} - V_1\right)\right]}{\partial \left(1/T\right)} , \tag{61}$$

where P_1^0 is the saturated vapor pressure of the sorbate, B_{11} is the second virial coefficient, and V_1 is the molar volume of the sorbate.

The fraction of the material in the transition zone may be calculated, in analogy with determining the crystallinity from inverse gel-permeation chromatography (IGC) data, according to the equation:

$$q = \frac{V_G(E) - V_G(B)}{V_G(E)} , \tag{62}$$

where $V_G(E)$ is the total value of the specific retention volume of sorbate by the first network and by the intermediate region, and $V_G(B)$ is the specific retention volume of sorbate by the first network only.

The enthalpy of mixing found from Eq. 61 and the fraction of the interphase region found from Eq. 62 are given in Fig. 15 [65]. It can be seen that at low concentrations of the second network ΔH_{12} is negative. This means that in this composition range there may be true miscibility. However, the enthalpy of mixing increases with concentration of the second network, and the free energy of mixing becomes positive at $w_2 = 0.3$.

The increase in ΔH above zero in Fig. 15 is accompanied by an increase in q, that is, by an increase in the size of the boundary region where thermodynamic incompatibility is observed between two evolved phases. Thus, a strictly thermodynamic quantity indicates the emergence of the phase boundary regions. The quantity of interfacial boundary materials may be increased partially at the expense of the amount of polymer II present. It should be noted that the interphase region arises as a result of the incomplete phase separation. Thus, its volume fraction should also depend on the conditions of phase separation that are determined by the kinetics of chemical reaction. The faster is the reaction of the network formation, the smaller should be the fraction and the thickness of the interphase.

The activation energies E_A of the IPN and of the homopolymer transitions were also calculated [94]. Analysis of the data indicates that the introduction of the second network (PS into the first PU) leads to a decrease in E_A in comparison with individual networks. This effect seems to be connected with the fact that the intermediate region has a loosely packed structure.

Another way to establish the existence of an interfacial region in microphase-separated IPNs is based on the application of differential scanning

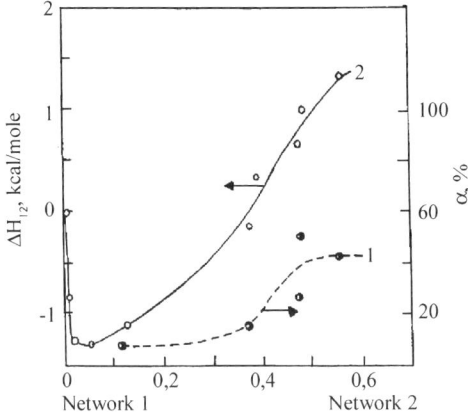

Fig. 15 Dependence on the fraction of transition layer α (1) and mixing enthalpy (2) on the content of the second network in the IPN [65]

calorimetry (DSC) [67, 96, 97]. It was found that the temperature dependence of heat capacity C_p is characterized by a broad curvature between two glass transition temperatures corresponding to separated phases. This is a consequence of the presence of an interfacial region which possesses a locally varying composition. The shape of the DSC trace contains substantial information about the character of an interphase. Hourston et al. [98–100] proposed a quantitative method for the determination of the weight fraction of an interphase and the extent of phase separation using a modulated DSC method. It is supposed that for fully immiscible blends where an interface exists between two phases, the value of the increment of heat capacity ΔC_p of any component diminishes in the blend owing to the transition of some part of material into the interfacial region. For this case the ΔC_p value of one component in the blend must be exactly the product of that in its pure state multiplied by its weight fraction, i.e., $\Delta C_p = \Delta C_p^0 \varphi$, where ΔC_p^0 is an increment for a pure component. For immiscible polymers the total ΔC_p is just the algebraic sum of the values of increments of two polymers. The calculations use the values of the increments of heat capacity ΔC_p at the corresponding glass transition temperatures, T_g, which decrease in the blend as compared with the pure component. The amount of a material in the interfacial region could be related to the value of F as defined below (Freed equation [101]):

$$F = \frac{\left(\varphi_{10}\Delta C_{p1} + \varphi_{20}\Delta C_{p2}\right)}{\left(\varphi_{10}\Delta C_{p10} + \varphi_{20}\Delta C_{p20}\right)}. \tag{63}$$

Here, ΔC_{pi0} values are the increments of heat capacity at T_g of components before mixing, ΔC_{pi} are increments in the blend, and φ_i are the weight fractions of components. These increments decrease when the proportion of two components is changed. Such behavior is attributed to the presence of a large amount of material in the interfacial region. The total increment of the heat capacity is equal to

$$\Delta C_p = \Delta C_{p1} + \Delta C_{p,\text{interphase}}, \tag{64}$$

where $\Delta C_{p1} = \varphi_1 \Delta C_{p10}$ and $\Delta C_{p2} = \varphi_2 \Delta C_{p20}$.

When no interface exists, the value of F is equal to 1.0. The weight fractions of polymers I and II in the interphase can be calculated from the following equations:

$$\delta_1 = \varphi_{10} - \frac{\Delta C_{p1}}{\Delta C_{p10}}, \delta_2 = \varphi_{20} - \frac{\Delta C_{p2}}{\Delta C_{p20}}. \tag{65}$$

Authors have also shown that values of dC_p/dT change with temperature at different phase separation times and, correspondingly, the fraction of interphase changes in the course of phase separation. The approach considered above may be applied, as follows, only to those systems that are separating in two pure phases. The application of the Freed equation to IPNs may be done if one has in mind its approximate character. The reason is that in IPNs the separated phases consist simultaneously of both components and each separated phase cannot be considered as an individual component. At the same time in evolved phases enriched with ions of one or other component, the content of the second polymer usually is very low and we believe that for comparative estimations the Freed equation may be used. Unfortunately, there are no more available physical methods to estimate the fraction of an interfacial region.

The interphase dimension may be estimated using both DSC and small-angle X-ray scattering (SAXS) [102, 103]. For sequential IPNs based on PU and poly(2-hydroxyethyl methacrylate) (PHEMA) it was observed that the interfacial thickness was zero, and there were sharp domain boundaries if they were estimated by SAXS. At the same time, according to DSC, a diffuse interphase region was observed. Thus, the results of the two methods are conflicting. For the IPNs based on PU and PS grafted via 2-hydroxyethyl methacrylate (HEMA) residues (1, 2, 5, 10 mass %), it was discovered that on increasing the cross-linking density of the first network, the weight fraction of the interphase increases [104]. The existence of an interfacial region follows from the calorimetric data. It was found by dividing the dC_p/dT signal into three parts related to PU-rich and PHEMA-rich phases and the interphase. The content of the latter was about 42% by mass.

Authors have made an attempt to estimate the interphase thickness. The method was based on SAXS data. The idea was the following. The specific interfacial area S/V was defined as the ratio the interfacial area S to the volume V of an interphase. Then

$$S/V = 4\varphi \left(1 - \varphi\right)/a_c , \qquad (66)$$

where a_c is the characteristic distance (the size of heterogeneity), which may be easily found from SAXS data. Let δ be the average interfacial thickness. Consider the case $\delta \ll V^{1/3}$; δS is equal to the volume of an interphase. The volume fraction of an interphase is $w = \delta S/V$. Then the thickness δ may be calculated as

$$\delta = wa_c/\left[4\varphi(1 - \varphi)\right] . \qquad (67)$$

This relation between the interfacial thickness of an interphase and correlation length could allow calculation of δ if the density of an interphase were known. However, the density of an interphase is not known.

2.3.5
Composition and Ratio of Phases Evolved During Phase Separation

From the discussion above it follows that in order to characterize the structure of IPNs with incomplete phase separation, it is very important to know the composition of each evolved phase and the composition of the interphase region. Up to now such data are not available because of experimental and theoretical difficulties. In principle, such estimation may be done in the following way. If we know the position of the glass transition points for each phase and there is no distinct maximum for an interphase, the composition of each phase may be calculated. For this purpose, one of the equations connecting the glass temperatures of components and glass transition temperature of the miscible polymer systems may be used. Such estimation was performed [67] using the well-known and the most simple Fox equation [96]:

$$1/T_g = \varphi_1/T_{g_1} + \varphi_2/T_{g_2}. \tag{68}$$

Here, φ_1 and φ_2 are the volume fractions of components, T_{g_1} and T_{g_2} are their glass transition temperatures, and T_g is the glass transition temperature of the miscible blend. Considering each phase as an independent quasi-equilibrium IPN and apparently one-phase system with forced miscibility (because it is characterized by its own glass transition), the composition and the ratio of phases may be found:

$$\varphi_1 = \frac{T_{g_1} T_{g_2} - T_{g_1} T_g'}{T_g' \left(T_{g_2} - T_{g_1}\right)} \tag{69}$$

and

$$\varphi_2 = \frac{T_{g_1} T_{g_2} - T_{g_2} T_g''}{T_g'' \left(T_{g_1} - T_{g_2}\right)}, \tag{70}$$

where T_g' and T_g'' are glass temperatures of evolved phases, and φ_1 and φ_2 are the volume fraction of each component in the phase enriched in this component.

It is clear, however, that if we prepare IPNs separately by taking as initial concentrations of components those that correspond to evolved phases, then as a result of curing such a system will again separate into two phases of different composition, because the conditions of phase separation for this system will be different from those for the initial IPN. This conclusion, in particular, follows from the nonequilibrium state of phase-separated IPNs.

A similar approach to estimating the composition of phase-separated regions and their content, based on the above-mentioned Fox equation, was later used for IPNs based on PU and maleimide-terminated PU [97], PU–epoxide–episulfide resin [102], and other systems. The results were not satisfactory, which can be explained by the approximate nature of the Fox equation (presently, there are a great number of equations connecting the glass transition temperature in compatible blends with transition temperatures of components).

2.4
Nonequilibrium States of IPNs

The microphase structure of IPNs may be described using the concept of the formation of various clusters: "physical clusters" that are formed due to phase separation of constituent networks under conditions of nonequilibrium phase transitions and "chemical clusters" due to cross-linking [51, 103]. As will be shown in Chap. 5, this volume, the most distinguished feature of simultaneous IPN formation consists of simultaneous proceeding of the formation reaction and the microphase separation. Both processes are superimposed. This leads to the close relation between the reaction kinetics, and the kinetics of phase separation.

Both the chemical reactions and the phase separation proceed under nonequilibrium conditions simply because they proceed simultaneously. After some degree of chemical conversion and cross-linking is reached, microphase separation is impeded and the system freezes in a nonequilibrium structure characterized by incomplete phase separation. Thus, by the completion of IPN formation, reactions proceed in two evolved phases. The real structure of an IPN is a multiphase one, which is determined by the coexistence of at least three "phases" (not in a true thermodynamic sense). Two phases are formed by networks due to phase separation. Each phase may be considered as an independent IPN in which phase separation did not take place (the state of "forced" compatibility), and in which mixing on the molecular level is preserved. The composition of these two phases is determined by the reaction rate and the temperature. Each phase has an average composition that does not correspond to the network ratio in the entire IPN. The third "phase" is the nonequilibrium transition zone from one phase to another; its size depends on the conditions of phase separation. This zone may be called mesophase and may be considered as a nonequilibrium IPN of some transition composition, since the molecular level of mixing should also be preserved. For spinodal decomposition there is no sharp border between coexisting phases. The transition zone may be arbitrarily chosen in such a way that its composition corresponds to the average composition of the IPN.

In general, the microphase structure of IPNs may be described as a nonequilibrium one. Indeed, if the phase separation were realized under equilibrium, then, in accordance with the most general thermodynamic rules, the composition and the ratio of phases would be determined only by the phase diagram of the system, and not by the conditions of the separation or chemical kinetics. The situation typical of IPNs and of linear polymer blends may never be realized in polymer solutions or in alloys of low molecular mass substances. Thus, the first reason for the nonequilibrium consists in the specific conditions of IPN formation.

However, in such nonequilibrium IPNs one can still discern various microregions that can be described as in quasi-equilibrium, that is, microregions with a near molecular level of mixing. Two thermodynamic states in the IPN can be distinguished. The IPN as a whole is a nonequilibrium system due to incomplete phase separation and thermodynamic immiscibility of the constituent networks. However, the two phases evolved may be considered as quasi-equilibrium phases because they are the result of microphase separation, and each phase preserves the composition that is almost the same as that of the one-phase state. This implies that the evolved phases are characterized by a molecular level of mixing, and, because the mixing level that was inherent to the state of mixing at earlier reaction stages is essentially preserved, which corresponds to the onset of phase separation, these phases are called quasi-equilibrium. Thus, the nonequilibrium microphase structure of IPNs may be presented as a microheterogeneous two-phase system that lacks molecular mixing of two constituent networks throughout the bulk, and which has a near molecular level of mixing in each phase and transition zone (interphase region).

All the ideas connected with the nonequilibrium multiphase structure of IPNs are in good agreement with a comparatively low segregation degree in IPNs (see Chap. 4, this volume). In such a way the whole structure of IPNs may be presented as a mesophase matrix with embedded microphase regions, which represent two evolved phases. Such a structural model coincides with the spinodal mechanism of decomposition.

Depending on the kinetic conditions of curing, various degrees of phase separation and various nonequilibrium states may be realized. To compare different systems, it is important to characterize these states by one parameter, χ_{AB}:

$$\chi_{AB} = Z \frac{\Delta w_{AB}}{kT} , \tag{71}$$

where Δw_{AB} is the exchange energy and Z is the coordination number of the lattice. For various nonequilibrium states of IPNs, the level of molecular interaction between different chains may be different. Formally, the difference may be ascribed to various values of Z. The system is in a state of true thermodynamic equilibrium only if the free energy of mixing or χ_{AB} is negative. However, for immiscible systems like IPNs the free energy of mixing and χ_{AB} are positive, as was discussed above. To compare various nonequilibrium states of IPNs, the interaction parameter may be found experimentally by various methods [52]. Now let us consider an IPN separated into two phases. Each quasi-equilibrium phase may be thought of as a solution of network A in network B and vice versa, that is, as a miscible system. If phase separation is complete, we deal with an equilibrium system that consists of two equilibrium phases with negative interaction parameters for each phase. In this case,

$$\left(\chi_{AB}\right)_{exp} = \left(\chi_{AB}\right)_{I} \varphi_{I} + \left(\chi_{AB}\right)_{II} \varphi_{II} , \tag{72}$$

where indices I and II relate to phases I and II and φ are their volume fraction, $\varphi_{I} + \varphi_{II} = 1$. If phase separation is not complete, $(\chi_{AB})_{exp}$ can be written as

$$\left(\chi_{AB}\right)_{exp} = \left(\chi_{AB}\right)_{I} \varphi_{I} + \left(\chi_{AB}\right)_{II} \varphi_{II} + \left(\chi_{AB}\right)_{m} \varphi_{m} . \tag{73}$$

In this equation, $(\chi_{AB})_m$ is the interaction parameter for the transition zone (interphase region) between the two phases, which determines the free energy of interphase formation; φ_m is the volume fraction of the interphase region. For immiscible systems, $(\chi_{AB})_{exp}$ is positive. Generally, the two regions of microphase separation are not in a state of true equilibrium and the system as a whole is also in a nonequilibrium state. Parameters $(\chi_{AB})_I$ and $(\varphi_{AB})_{II}$ may be either positive or negative, whereas $(\chi_{AB})_m$ should always be positive. According to Helfand [104], $(\chi_{AB})_m$ characterizes the interaction at the interface between two phases (in the interphase region). If the transition region is formed as a result of incomplete phase separation (see below), then this nonequilibrium region of uncompleted separation may preserve the one-phase structure of initial (before phase separation) mixture. In this case the forced compatibility is realized and parameter $(\chi_{AB})_m$ may be negative.

As is seen from the last equation, the value of χ_{AB} consists of three parts, each being determined by the conditions of phase separation and by the deviation of the system from equilibrium. The experimentally found interaction parameter is determined by the system history. It has the physical meaning of characteristics of the system nonequilibrium degree. It also allows us to compare systems under identical conditions, but with different paths leading to these conditions.

In conclusion, one has to say that the general state of the investigation of the thermodynamics of IPN mixing and phase separation is far from the one we have for blends of linear polymers [52]. In particular, one very important factor is not accounted for by studying the miscibility of two networks, namely, the changes of volume by mixing. The last parameter is important not only for thermodynamics, but also for the structure and viscoelastic behavior of IPNs and is considered below.

3
Heterogeneous Structure and Morphology of IPNs

Phase-separated IPNs are heterogeneous systems characterized by the presence of two phases of various compositions, with the transition layer between them arising as a result of the incomplete separation. It is well established [23] that the usual cross-linked polymers also possess heterogeneous structure due to the mechanism of gel formation. In IPNs this heterogeneity of each constituent network should superimpose on the microheterogeneity due to phase separation in the course of IPN formation. It is very important to characterize the chemical and molecular and supermolecular structure of IPNs as well as their heterogeneity, which is tightly connected with the degree of phase separation (segregation degree). It is these characteristics that determine the viscoelastic and mechanical properties of IPNs. The chemical structure of IPNs, being determined by the chemical composition of the constituent networks, can be characterized by the cross-linking degree, whereas the molecular structure is characterized by the free volume. The physical structure of IPNs may be found either by X-ray scattering methods, or by electron microscopy. The difficulties connected with applying SAXS and SANS structural methods for investigation of amorphous polymers, including networks, are very well known, and because of this there exists not so much data on the IPN structure, based on these methods. At the same time, there are a large number of published papers dedicated to the morphology of IPNs as studied via electron microscopy. However, while the reliability of the scattering methods is evident, electron microscopy cannot give any quantitative evaluations of the structure and connect them with physical properties. Indeed, electron microscopy is more an art than a science, because the interpretation of results for amorphous polymers, as a rule, is rather vague.

3.1
Cross-linking Density in IPNs

In spite of many investigations of the structure and properties of IPNs, the questions connected to the network density in IPNs still have no answers. Meanwhile, the effective network density is one of the most important characteristics of any network, including IPNs. To study IPN properties, first of all it is necessary to compare its density with the densities of the constituent networks. To calculate the effective cross-linking density it is necessary to know the usual physical densities of both networks, which are used in the following calculations.

Usually two methods of estimating the effective cross-linking density are used. The first is based on elasticity theory. It computes the concentration of the cross-links in the network from the value of the elastic modulus and the polymer density according to the equation:

$$\nu_e = \frac{E}{3\rho RT} , \tag{74}$$

where ν_e is the effective concentration of the cross-links, E is the elastic modulus, and ρ is the network density [105, 106]. The equation was derived from the following assumptions: the chains forming the network have equal full length, the distribution of the chain ends obeys the Gaussian statistics, the volume is not changed by deformation, and the deformation is affine. The serious limitations in using this equation, even for rubberlike polymers, can be easily seen. This equation was derived for vulcanized networks and did not take into account possible chain entanglements.

Meanwhile, chain entanglements in IPNs play a very important role in the processes of their microphase separation, by preventing it, which early has not been taken into account. Nevertheless, this method was used in many works. One example is the study of the sequential IPNs based on copolymers of styrene and DVB [107] for various ratios of constituent networks w_1 and w_2. The value of ν_e was compared with the density of the cross-links in networks ν_r, calculated from the content of corresponding monomers in primary (w_1) and secondary (w_2) networks. It was found that the dependence of ν_e on ν_r is linear; however, the values of ν_e for the IPN were higher as compared to the primary network. This effect was explained by the contribution of the mechanical entanglements between networks.

The second method of computing the cross-linking density is based on the well-known Flory–Huggins theory, and was initially used in some works by Frisch, Frisch, and Klempner [9, 108, 109].

To determine ν_e of a homogeneous (swollen) polymer network, the equilibrium modulus is measured in shear, compression, or extension [31, 111]. The stress in uniaxial extension/compression mode is equal to

$$\sigma = \frac{f}{A_{sw}} = RTA_f \nu_e \varphi_2^{1/3} \varphi_0^{2/3} \left(\Lambda - \Lambda^{-2} \right) . \tag{75}$$

In this equation A_{sw} is the cross section of the swollen specimen, R is the gas constant, T is temperature in Kelvin, φ_2 is the volume fraction of polymer in the swollen sample, φ_0 is the fraction of polymer-building components during network formation, and Λ is the deformation ratio. A_f is the front factor (see [111]). The value of ν_e refers to unit unswollen volume.

It was suggested that if ν_e is higher for IPNs compared to the average value calculated from the ratio of networks, the networks are interpenetrating. The following important assumption, later confirmed experimentally [112], was made. The IPN consists of three regions: network 1, network 2, and the intermediate region (interphase). In such a way from the very beginning it was accepted that there is no pure interpenetration of the chains of both networks. The total effective cross-linking density was considered to be

$$\nu_e = \Phi_1 \left(\nu_e \right)_1 + \Phi \left(\nu_e \right)_2 + \Phi_3 \left[\left(\nu_e \right)_1 + \left(\nu_e \right)_2 \right] , \tag{76}$$

where $(\nu_e)_1$ and $(\nu_e)_2$ are the densities of the individual networks, Φ_1 is the volume fraction of the i-th region, $(\Phi_1 + \Phi_2 + \Phi_3) = 1$. If $\Phi_1 = \Phi_2$, we have

$$\nu_e = \left(\frac{1}{2} \right) \left[\left(\nu_e \right)_1 + \left(\nu_e \right)_2 \right] \left(1 + \Phi_3 \right) . \tag{77}$$

Because of this, the value Φ_3 is

$$\Phi_3 = \frac{\nu_e - \left(\nu_e \right)_0}{\left(\nu_e \right)_0} , \tag{78}$$

where $(\nu_e)_0$ is the cross-linking density of the system with no interpenetration.

Such consideration supposes that the interpenetration proceeds only in one part of the system, in accordance with the two-phase structure of IPNs. It was assumed that two phases interpenetrate only in the intermediate regions between two phases. However, the concept of interphase was introduced formally with the sole purpose of describing the deviation of the network density from additivity, and the value of φ_a does not really give any evidence of the existence of the transition zone.

In spite of these limitations, the application of this method has produced some interesting results. For example, for the IPN based on PU and polyacrylate it was found that there exists a strong dependence of the cross-linking density on composition, the absolute values being much higher as compared with those calculated from the stoichiometric composition of the IPN. For sequential IPNs based on PU and styrene–DVB copolymer, it was found that the experimental values of the cross-linking density had intermediate values between the densities of constituent networks. In some cases the theoretical values of v_e exceed the experimental ones at some composition and are simultaneously lower at some other composition.

For semi- and full IPNs made of poly(oxyethylene) and poly(acrylic acid), the effective cross-link densities were determined from the elastic modulus and were compared with values estimated assuming the additivity of cross-link densities of components [113]. Discrepancies between estimated and calculated values are observed.

One of the most widely distributed methods of evaluating the effective cross-linking density is based on the data on equilibrium swelling. For this purpose the Flory–Rehner equation is used [105]:

$$- \left[\ln \left(1 - v_2 \right) - v_2 + \chi_1 v_2^2 \right] = v_1 \left(v_e / V \right) v_2^{1/3} - 2 v_2 / 3 , \tag{79}$$

where $v_e / V = \rho / M_c$ is the effective network density, ρ is the polymer density, M_c is the molecular mass of segments between cross-links, v_2 is a volume fraction of polymers in a swollen gel, v_1 is a molar volume of a solvent, and χ_1 is the interaction parameter.

This parameter is usually calculated from the data on swelling obtained at various temperatures. The main problem in using this equation for IPNs is to find such a solvent that has similar values of the interaction parameter for both polymers. Experimental data have shown that in almost all cases the values v_e / V are higher as compared with theoretically found values, this probably being the result of the entanglements of the chains.

Thiele and Cohen [114] derived an equilibrium swelling equation for homo-PS/PS IPNs where both networks are chemically identical except for cross-linking level. This equation was modified by Siegfried, Thomas, and Sperling [115] and describes the equilibrium swelling of sequential IPNs with a single T_g, i.e., for a one-phase system. It is evident that this equation cannot be transferred to phase-separated IPNs without considering the possibility of applying this theory to each evolved phase. We believe that if the ratio and composition of two phases were known, it would be possible to estimate the total cross-linking density due to entanglements using the additivity rule. The modified equation for equilibrium swelling has the following form [115]:

$$\ln\left(1 - v_1 - v_2\right) + v_1 + v_2 + \chi_S\left(v_1 + v_2\right)^2$$
$$= - V_S N_1'\left(1/V_1^0\right)^{2/3}\left(v_1^{1/3} - v_1/2\right) - V_S N_2'\left[\left(V_2^0\right)^{2/3} v_2^{1/3} - V_2/2\right], \qquad (80)$$

where subscripts S refers to solvent, v_1 and v_2 represent the volume fractions of the two polymers in the equilibrium swollen state, v_1^0 and v_2^0 represent the volume fractions of the two polymers in the dry state, χ_S denotes the polymer–solvent interaction parameter (presumed identical for both polymers), V_S represents the solvent molar volume, and N_1' and N_2' represent the cross-links of the corresponding single networks in $mol \cdot cm^{-3}$ as determined by the Flory–Rehner equation. Experimentally this equation was used to analyze PS/PS homo-IPNs, which were synthesized by swelling cross-linked polymer I with monomer II plus cross-linking and activation agents, and polymerizing II in situ.

In the derivation of Eq. 80 the contribution of mutual physical entanglements was not considered. If there are some physical cross-links arising from internetwork entanglements, the swelling data are expected to be shifted on the plot of the theoretical (calculated from stoichiometry) and the experimental values of cross-linking density. The effective cross-link density N, calculated from the Young's modulus, usually exceeded theoretical values calculated from stoichiometry. It is taken as evidence for the new physical cross-links caused by entanglements formed by IPN formation.

As the cross-links density of network 1 is increased, the shift is more pronounced, indicating the existence of mutual entanglements. As is seen this calculation allows us only to establish the existence of mutual entanglements, but gives no possibility to estimate in a quantitative way their contribution to the total cross-linking density.

In many investigations of cross-linking densities for phase-separated IPNs, authors accept the additive scheme and calculate the IPN density from corresponding stoichiometry calculations [116]. However, due to special features of the formation of one network in the presence of another, the additive rule is generally not valid. At that time, there were no attempts to develop a theory of cross-linking in IPNs that will take into account the cross-linking densities in separated phases as well as between them. Using all the equations describing the swelling of the network, one has to bear in mind the fact that the dependence of the swelling degree on the nature of polymer and solvent cannot account for entanglements.

Later, by considering the filler influence [117] on the network density it was shown that filled IPNs have lower network density as compared with unfilled, due to filler influence on the reaction conditions.

In comparison with traditional cross-linked polymers, in IPNs, due to the principle of their formation, there should be a strong contribution of chain entanglements into the effective network density. The effective network density consists of two terms:

$$\nu_{eff} = \nu_{chem} + \nu_{ent} , \tag{81}$$

where ν_{chem} is determined by the stoichiometry of reaction and ν_{ent} by the conditions of IPN formation and microphase separation by cross-linking. For an ideal IPN, ν_{eff} should always be higher due to the entanglement contribution. In real IPNs, ν_{chem} can be different from the sum of network densities of constituent networks. However, values of ν_{eff} that are lower than the additive value show that the network structure is full of defects and the entanglement contribution does not compensate for the loss in density, since ν_{eff} depends on the reaction conditions and on the mutual influence of both decomposing networks on the total cross-linking degree. This effect is described in [118, 119]. It is connected to the influence of the reaction kinetics on the phase separation. During fast reaction there is enough time for phase separation, and the entanglement contribution should be the highest one. On the contrary, for low reaction rates, the microphase separation is ended before the gel point, and the role of entanglement should be low. Besides, one has to bear in mind that the final formation of the IPN structure proceeds in two separate phases, in each of which the reaction continues up to the gel point. The total cross-linking density should take into account the cross-linking density in each reacting phase, and for this purpose it is necessary to know the phase ratio. Practically, this task is too complicated to be solved experimentally with the present methods. Since IPNs are phase-separated systems, one has to discuss the cross-linking density in each phase and in the interphase, which is not possible yet.

A conclusion can be drawn that to determine the effective cross-linking density, the methods based on the modulus measurements should be preferred. It seems possible in this case to distinguish between two contributions to the ν_{eff}, if ν_e is determined chemically. An attempt to establish the interconnection between effective cross-linking density in semi-IPNs and kinetic conditions of reaction has been made for IPNs based on cross-linked PU and linear or partially cross-linked PBMA.

Analyzing the experimental data on swelling in IPNs, as was already mentioned, is very difficult. There are no theoretical descriptions of this process either for sequential or simultaneous IPNs, because in all cases in phase-separated systems there exist two phases, each being an IPN of different composition. In our approach we consider the swelling on the basis of the following concept. Semi- and full IPNs are phase-separated systems formed of two phases enriched in one of the IPN components. These phases may be considered as nonequilibrium networks with a molecular level of mixing in each phase [103]. For semi-IPNs that means that such an "ideal" phase is formed by a network polymer and a linear or branched one distributed in the main network cell between cross-links. Swelling of such an ideal network depends only on the cross-linking density of the network, as the linear or branched polymer in the network cell is restricted in its ability to swell and does not depend on the swelling ability of the linear component. Such a situation should exist for both phases evolved during reaction and microphase separation. The experimentally found swelling degree is the sum of swellings in both phases. Because these contributions cannot be separated, we shall operate further with the total swelling effect.

To establish the kinetic influence on the effective cross-linking density in semi-IPNs, the kinetics of the formation of a pure PU network and of polymerization of butyl methacrylate (BMA) have been investigated. It was found that with decreasing molecular mass of polydiol used for PU formation, the rate of network formation increases together with conversion degree at the gel point (from 0.60 to 0.77). Kinetic data show that the reaction kinetics is interconnected. Values of M_c calculated from the swelling data are presented in Table 2 [119].

It can be seen that for PU based on poly(oxypropylene glycol) (POPG)-500 and POPG-1000, experimental values of M_c are lower as compared with theoretical ones calculated from the reaction stoichiometry, whereas for POPG-1500 and POPG-2000 the experimental data are higher. The higher cross-linking density for diol, whose molecular mass is low, was explained by the contribution of strong physical bonds to cross-linking density. The transition from linear PU to semi-IPN is accompanied by the increasing equilibrium swelling of the system. For the IPN based on POPG-500, the swelling degree increases four times as compared with pure PU, and for PU based on POPG-2000 it increases 1.4 times. Therefore, simultaneous formation of both cross-linked and linear PBMA leads to the formation of a network structure with a higher number of defects as compared with other systems under investigation. This is probably connected with the fact that the rate of PBMA formation in the semi-IPN based on POPG-500 is one order higher as compared with POPG-2000. The formation of the network structure of PU on

Table 2 Experimental values of M_c (exp) and theoretically calculated M_c (theor) for PU and semi-IPNs[a] [119]

System	M, polydiol	M_c, exp	(ν_e/V) exp 10^4 mol·cm^{-3}	M_c, theor	(ν_e/V) theor, 10^4 mol·cm^{-3}
PU	2000	3600	3.00	2380	4.54
PU	1500	2740	3.95	1880	5.77
PU	1000	1200	9.20	1380	8.07
PU	500	450	25.70	880	13.20
semi-IPN	2000	4900	2.10	880	13.20
semi-IPN	1500	3800	2.73	–	–
semi-IPN	1000	3050	3.53	–	–
semi-IPN	500	1840	5.80	–	–

[a] Theoretical values for PU and semi-IPNs are the same

POPG-500 proceeds not in a liquid medium, as for PU with POPG-2000, but in a medium with high viscosity (due to strong intermolecular interactions). As a result, a less regular structure of the PU network is formed and we observe weaker effective cross-links as compared with the pure PU network.

The formation of semi-IPNs based on POPG with lower molecular mass proceeds faster under conditions preventing phase separation. Under such conditions the contribution of entanglements to the effective cross-linking density should be more pronounced. However, if we compare the ratio M_c (semi-IPN)/M_c (PU), we can see that the lower is the molecular mass of POPG, the higher is this ratio. This means that the cross-linking density is less, the lower the MM of POPG.

Now, if we apply the model approach proposed above, we shall see that the main effect on the effective cross-linking density in semi-IPNs is produced by the kinetic conditions and the nature of the components. Changing reaction conditions leads to the formation of the more defective structures of networks as compared with the pure network. Thus, the effective network density is determined not only by the theoretical topology of the network but also by the reaction conditions. It is worth noting that the formation of IPNs under phase-separated conditions enhances the formation of the defective network because phase separation leads to the formation of two phases and of a transitional region, which may be considered as an independent IPN with its own composition and cross-linking density.

In such a way we see that the problem of determining the cross-linking density in IPNs is far from a solution, in spite of its importance for characterizing the mechanical properties of advanced materials. At the same time one has to bear in mind that, due to the conditions of their formation (kinetics, phase-separation, etc.), one and the same IPN may be characterized by various values of cross-linking density; therefore, this parameter cannot serve as determining their mechanical and other properties. The higher is the phase separation, the smaller is the contribution of entanglements to the effective network density. We believe that this uncertainty is the reason for the fact that in all investigations of IPNs dedicated to their relaxation and viscoelastic properties, the qualitative data on cross-linking degree are absent, and the only item that characterizes these properties is the reaction stoichiometry.

As often happens in science, instead of making theoretical and experimental attempts to elucidate the problem of determining the cross-linking density, this problem was simply forgotten. Even in the famous book by de Gennes [47] there is no mention about the cross-linking density of networks.

3.2
Free Volume

The appearance of two phases in IPNs during curing and the existence of the interphase show that the packing density of IPNs is looser as compared with pure constituent networks. The transition of polymer chains into the interphase is associated with a gain in entropy due to decreasing packing density in this region, consequently leading to an increased free volume [120].

For polymer blends it was shown [121] that in the interphase region, the fraction of free volume increases compared with coexisting phases. The higher is the fraction of the interphase, the larger is the contribution of the free volume in the interphase to the total free volume of the system.

The free volume in IPNs has not been investigated extensively. The first attempt to estimate this value was done [65] using the data on vapor sorption based on the theory developed by Fujita [122]. The sequential IPNs based on styrene–DVB copolymer (network I) and cross-linked PU (network II) were studied. The PU content was up to 0.24 by mass. From the data on the sorption kinetics the interdiffusion coefficients D_v, solvent self-diffusion coefficient D^*, and relative diffusion coefficient D were found. The value of D was calculated from the relation $D_v = D (1 - v_s)$, where v_s is the volume fraction of a solvent in the IPN. According to Fujita, the change in self-diffusion coefficient in isothermal conditions is described by the equation:

$$\ln \frac{D^*}{D(0,T)} = \frac{B}{2.3 f^2 + f(T) \beta(T) \varphi_s} \frac{\beta(T) \varphi_s}{}, \tag{82}$$

where $f(T)$ is the fraction of free volume of a polymer at T, $D(0,T)$ is the relative diffusion coefficient at $\varphi_s = 0$ and temperature T, B is a constant, and $\beta(T) = f_s(T) - f_p(T)$ is a parameter characterizing the solvent contribution to an increase in the free volume of a system. Here, f_s and f_p are the fractions of free volume of a solvent and polymer, respectively. The experimental data were plotted as

$$\varphi_s \left[\log \frac{D^*}{D(0,T)} \right] = \frac{2.3 f^2(T)}{B\beta} + \frac{2.3 f(T)}{B} \varphi_s . \tag{83}$$

The dependence of $\varphi_s[\log(D^*/D(0,T)] - 1$ on φ_s is linear, which indicates the applicability of the free volume model to the system under investigation. From these data the values of $f_p(T)$ were found from known values of f_s. The data for various ratios of networks were compared with those calculated according to the additivity rule (Fig. 16) [65].

As we see, values of Δf change in a nonmonotonous way depending on the network ratio, the fraction of free volume being higher in IPNs at all ratios but $w_2/w_1 = 0.03$–0.07. The complicated dependence of the free volume on composition may be attributed to the formation of the intermediate region of the interphase between two phases in the IPN. A looser interphase contributes to the increasing free volume of the whole system.

The same Fujita method was used to estimate the dependence of the fractional free volume on composition for two IPNs based on PU and on ionomeric PU [123, 124]. The PU network was obtained from propy-

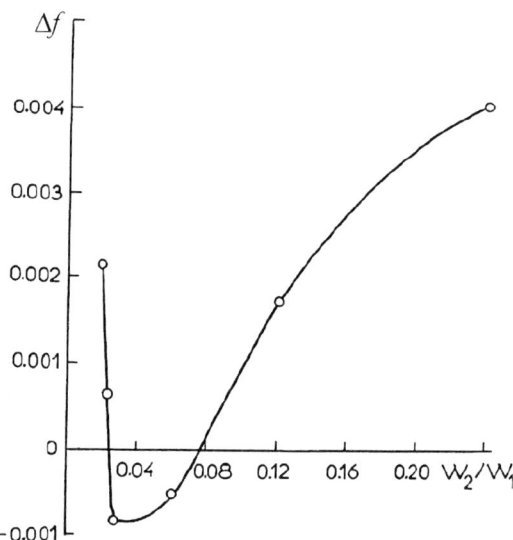

Fig. 16 Change of the fractional free volume deviation, Δf, found experimentally from additivity for various network ratios [122]

lene oxide–tetrahydrofuran copolymer and toluene diisocyanate (TDI)–trimethylolpropane adduct with the ratio of NCO:OH being equal to 2. PU ionomer was synthesized from cation-active oligomers, which were the products of the reaction between branched polypropylene glycol containing three OH side groups per molecule, 2,4-TDI, and 2,2-dimethylethanolamine. The three-dimensional network was created through the formation of ionic bonds by adding 1,5-dibromopentene to the oligomeric system. IPNs were obtained by simultaneous curing. Another system was made from the same PU and a linear ionomer (B) from polytetramethylene glycol, TDI, and the K salt of dioxybenzoic acid. The results of calculations are presented in Fig. 17 [124]. As one can see, for a small amount of cation-active ionomer (A) in the IPN, the value $f(T)$ is less than the additive one, whereas in the region of intermediate composition, IPNs are characterized by increased free volume fraction. For anion-active ionomer (B) the free volume fractions deviate from the additive ones over the entire concentration range. The extremes on the $f(T)$ vs composition curve coincide with those on the curve representing the dependence of the free energy of mixing of the same IPNs. From the comparison of the corresponding data it follows that higher compatibility is characterized by the decreased free volume fraction as compared with free components, whereas for the region of immiscibility the excess free volume is typical. It may be suggested that the self-association in the cation-active ionomer and segregation of components of the IPN are smaller than in anion-active ionomer, which is more capable of self-association. Simultaneously, ionomer A has three-dimensional structure, while ionomer B is a linear poly-

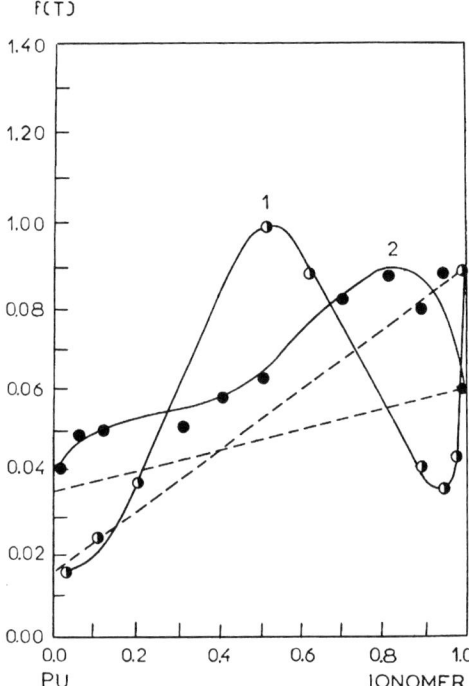

Fig. 17 Dependence of the free volume fraction $f(T)$ on the component ratio for IPNs containing ionomer A (curve 1) and ionomer B (curve 2); *dotted lines* show additive values [125]

mer. The segregation in the second case may be more pronounced, leading to the appearance of loosely packed interphase regions.

The fraction of the interphase and its change depending on the IPN composition could be calculated as follows. The free volume f of the system may be presented as:

$$f = (f_1 w_1 + f_2 w_2)(1 - a) + a(\varphi_1 w_{i1} + \varphi_2 w_{i2}) , \tag{84}$$

where f_1 and f_2 are the fractional free volumes of the first and second networks, φ_1 and φ_2 are the fractional free volumes of the first and second networks in the interphase, α is the fraction of the interphase, and w_{i1} and w_{i2} are mass fractions of the first and second networks in the interphase. The deviation of the fractional free volume from additivity is

$$\Delta f = a\left[(\varphi_1 w_{i1} + \varphi_2 w_{i2}) - (f_1 w_1 + f_2 w_{i2})\right] . \tag{85}$$

Because the surface tensions of IPN components are very close, the selective adsorption at the interface between two components is almost absent. Then the ratio of the networks in the interphase may be assumed to be the same as in the bulk of the IPN, i.e., $w_{i1} = w_1$ and $w_{i2} = w_2$. In this case, Eq. 85 may be written as:

$$\Delta f = a \left(\varphi_1 w_1 + \varphi_2 - \varphi_2 w_1 - f_1 w_1 - f_2 + f_2 w_1 \right)$$
$$= a \left[(\varphi_2 - f_2) + w_1 (\varphi_1 - f_1 - \varphi_2 + f_2) \right] . \tag{86}$$

This dependence may be transformed into

$$\Delta f = a(k_1 + k_2 w_1), \tag{87}$$

where $k_1 = \phi_2 - f_2$ and $k_2 = \phi_1 - f_1 - \phi_2 + f_2$.

The experimental dependence of Δf on w_1 for IPNs made from PU and styrene–DVB copolymer (Fig. 16) may be approximated by the equation:

$$\Delta f = e^{bw_1} \left(k_1 + k_2 w_1 \right) . \tag{88}$$

From the calculated parameters of Eq. 88 the fraction of the interphase was found. This value increases linearly with the content of the second network. This calculation implies that the deviations of the free volume from additivity are determined only by the formation of an interphase between two separated phases in the IPN. This allows the conclusion to be drawn that the shift of the temperatures of relaxation transitions in IPNs is not the result of some compatibility at the interface, but on the contrary, the result of incompatibility. Unfortunately, in the cited work the range of the network ratios studied was too small.

One can expect that the distribution of the free volume in IPNs should also differ from the ones in constituent networks. Based on the lattice model of free volume [125] the approach to the estimation of the free volume distribution in semi-IPNs was proposed [126]. It was accepted that free volume was a set of vacancies in lattices of various unitary dimensions inserted into one another or densely joined with one another.

To find experimentally the free volume distribution, a variant of gel chromatography was used that allowed estimation of the dimensions of molecules dissolved in eluent from the retention time t_R in gel. In [126, 127], inverse gel-permeation chromatography was used and the dimensions of free space or vacancies in the network were determined from the retention time of test molecules in the gel. These molecules were of various sizes up to monomeric ones. In such a way, the part of space between the cross-links in the network was found, available for permeation of molecules (macromolecules) with known mean-square distance between the chain ends, $< h >$. The larger the macromolecule dimension, the lower the total fraction of vacancies available for their penetration. Thus the distribution of vacancy size can be estimated. The main problem in the application of this method consists in the transition from the distribution obtained for swollen gel (the necessary condition for gel chromatography) to distribution for pure network without solvent.

Using the proposed method, the free volume distribution was determined for simultaneous semi-IPNs based on cross-linked PU and PBMA (simultaneous curing of PU and polymerization of BMA) [126]. The PU/PBMA ratio was 1 : 1. The gel chromatographic experiments were performed using as a test molecule the standard polystyrene (Du Pont) with MM from 1 800 000 to 800 at $M_w/M_n < 1.05$, toluene and styrene being also used as low molecular mass substances. The transition from the molecular masses to the mean-square dimensions of the coils for PS were done using the relation $< h > = 0.065\ M^{1/2}$.

From the dependencies of the retention time on the molecular mass of test molecules, the integral distributions of vacancies were found for cross-linked PU and semi-IPNs and the differential distribution of vacancies was calculated (Figs. 18, 19) [126]. For a pure network the greatest fraction of vacancies has the dimension of 1.32–1.75 nm. The dimensions below 1.32 nm are not present in the network. For semi-IPNs the vacancies above 14.5 nm disappear; the smallest vacancies had a broader distribution with a broad maximum at 1.24–2.00 nm. The average dimension of vacancies was calculated according to the equation:

$$\langle h \rangle_{av} = \frac{\sum_i \langle h \rangle_i \left(dI/d \langle h \rangle \right)_i}{\sum_i \left(dI/d \langle h \rangle \right)_i}, \tag{89}$$

where I is the integral distribution and i is the order of the derivative.

The comparison of the data for swollen PU and the semi-IPN shows that the average dimensions of the vacancies are rather close (1.80 and 1.87 nm,

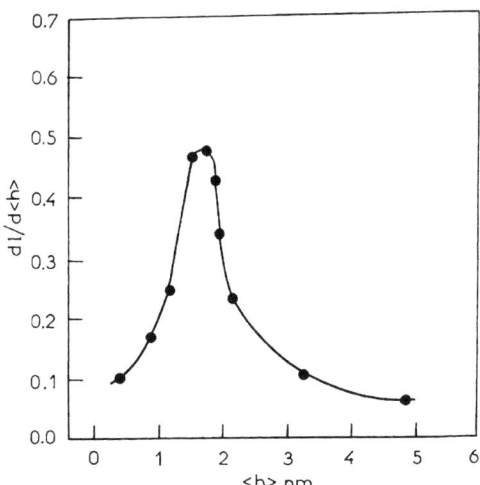

Fig. 18 Differential distribution of vacancies in PU [127]

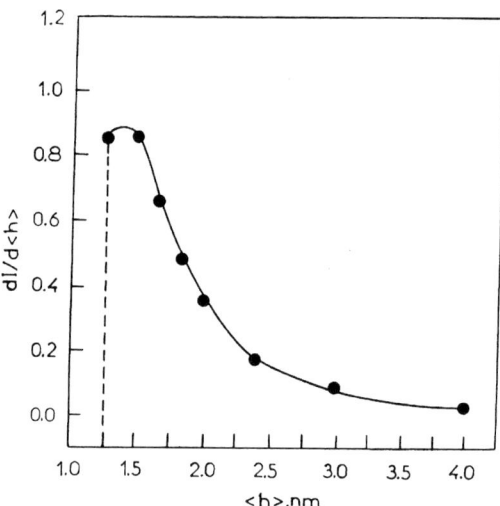

Fig. 19 Differential distribution of vacancies in the semi-IPN [127]

respectively), whereas their distributions differ significantly. The semi-IPN is more heterogeneous with respect to vacancy distribution and does not contain large vacancies. The fraction of free volume for small and intermediate dimensions in the IPN is higher as compared with PU, in spite of the presence in PU of large vacancies. The total fraction of free volume is higher for semi-IPNs. The total volume of vacancies may be considered as the free volume of the system. Its value was found for pure PU and semi-IPNs in the swollen state.

It is possible that the effects observed are connected to the network densities in cross-linked PU and the semi-IPN. To clarify the situation, the estimation of the effective network density of PU and the semi-IPN was done using the data on swelling. The interaction parameter χ_1 was calculated using the principle of corresponding states. The values of M_c were found to be 2400 for the PU network (the one calculated from stoichiometry) and 1700 for the semi-IPN. As we see, both the PU and semi-IPN have approximately the same network densities. As follows from these data, the higher fraction of free volume in the semi-IPN cannot be attributed to different degrees of cross-linking. The only explanation seems to be reasonable, namely, that in semi-IPNs, as in nonequilibrium systems, the majority of components are formed in an interphase region of incomplete phase separation. The diminishing packing density in this region may serve as a source of additional free volume in semi-IPN as compared to the pure PU network. Having found the values characterizing the network density, one can evaluate its influence on free volume distribution. It may be assumed that the distribution of free volume should be bimodal because of the presence of cross-links. Some part of the free volume should be associated with the network density, whereas the other part is associated with the free volume determined by chains between cross-links. The increment of free volume, Δf, was calculated as a function of $<h>$ for each kind of test molecule. Figures 20 and 21 show the corresponding dependencies. It is seen that bimodality really exists. The hole dimensions in the range 2.5–3.0 nm (part of the curve before increment begins to grow) correlate with values of M_c for PU and semi-IPNs. As was said, these data were obtained for the swollen networks using the gel chromatography principle.

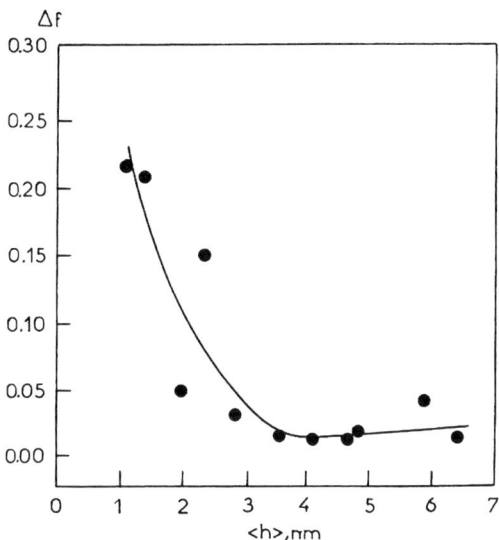

Fig. 20 Dependence of the increment of fractional free volume on cell dimensions for a PU network [127]

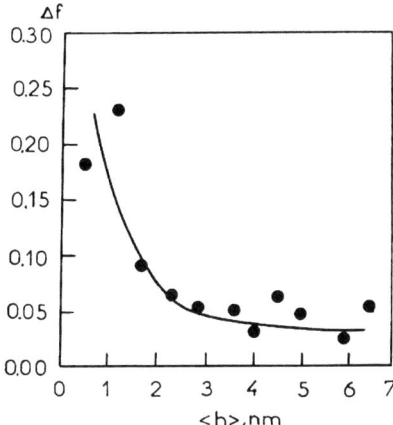

Fig. 21 Dependence of the increment of fractional free volume on cell dimensions for a semi-IPN [127]

To realize the transition from the vacancy distribution in swollen gel to the pure network, the following approach was proposed [126]. The system under investigation was assumed to be an affine network in accordance with the lattice model, where all cells are equivalent. Then, the calculation does not depend on the network type (semi- or full IPNs).

The volume of a nonswollen specimen V_p may be represented as consisting of two parts: free volume V_p^f and occupied volume V_p^0:

$$V_p = V_p^f + V_p^0 . \tag{90}$$

It is assumed that the swelling proceeds due to an increase in free volume. In such a case, if the free volume of the swollen specimen is V_s^f, we have:

$$sV_p = V_p + V_s^f - V_p^f , \tag{91}$$

where s is the degree of swelling and $V_s^f - V_p^f$ is an increment of free volume due to swelling. Solving the last equation for V_p^f, we obtain:

$$V_p^f = V_p + V_s^f - sV_p . \tag{92}$$

After dividing by V_p, the fractional free volume of the nonswollen specimen may be found as

$$f_p = (V_p + V_s^f - sV_p)/V_p . \tag{93}$$

Dividing both the numerator and the denominator of the right part of the equation by sV_p, we find f_p as a function of the experimentally found free volume fraction, f, of swollen specimen:

$$f_p = 1 - s(1 - f) . \tag{94}$$

This equation satisfies limiting conditions: for an unswollen specimen ($s = 1$) we have $f_p = f$ and at $f = 1$ (hypothetical case) $f_p = 1$, confirming the correctness of using Eq. 91 for deriving Eq. 94.

The corresponding values for nonswollen networks were calculated according to Eq. 94. The total fractions of the free volume are equal to 0.13 for PU and 0.46 for semi-IPN. Unfortunately, the restricted data do not allow us to make general conclusions about the contribution of free volume changes to the viscoelastic properties of IPNs. Experimentally, free volume in PU/polyacrylonitrile IPNs has been studied by using the method of positron annihilation [128].

3.3
Parameters of Heterogeneous Structure of IPNs

To evaluate the heterogeneity of the phase-separated IPN and the degree of phase separation (segregation degree) various approaches can be used. The most fundamental one is based on SAXS in the form proposed by Bonart and

Mueller [129, 130]. The following characteristics are important: the difference of the local density $\rho'(x)$ from the mean density ρ' and the value $\Delta\rho_c^2$, which represents an overall measure of all density variations. For multicomponent systems it is of great interest to compare the value of $\Delta\rho^{2'}$ with theoretical values of the mean square of electron density fluctuations, $\Delta\rho_c^2$. The latter value may be calculated from the electron phase densities ρ_1 and ρ_2 and from the phase volume fraction φ.

In this case:

$$\Delta\rho_c^2 = \varphi\left(1 - \varphi\right)\left(\rho_1 - \rho_2\right)^2 \tag{95}$$

$$\alpha = \frac{\Delta\rho^{2'}}{\Delta\rho_c^2}. \tag{96}$$

This ratio provides an overall measure for the degree of segregation.

It should be noted from the very beginning that segregation degree is a relative value, which characterizes the hypothetical system where α is the fraction of fully separated material whereas $(1 - \alpha)$ remains in the unseparated state. It is evident that the real situation does not correspond to such a simple picture. In spite of this, the value of segregation degree is a quantitative value which allows one to "probe" the most important properties of phase-separated IPNs.

The SAXS data also allow us to evaluate other very important characteristics of the heterogeneous structure. From these data the mean square of electron density fluctuation, $\Delta\rho^{2''}$, may be obtained. This value is calculated using the experimentally determined value of the thickness of the interphase layer E, which is also a very important quantity characterizing the transition region between two separated phases. The method of calculating E is rather complicated and is described in detail [130]. The boundary diffuseness β may be expressed as

$$\beta = 1 - \Delta\rho^{2''}/\Delta\rho^{2'}, \tag{97}$$

where $\Delta\rho^{2''}$ is the average square of the electron density fluctuations for the similar system, but not accounting for the effect of the diffuse interphase layer. Value of β does not depend on the nature of intermixed segments.

The next step of characterization is the estimation of the value

$$\gamma = 1 - \Delta\rho^{2''}/\Delta\rho_C^2, \tag{98}$$

which represents a measure of the number of intermixed segments, irrespective of any boundary diffuseness.

For the model of heterogeneous structure with a definite intermixing of segments, the parameter $\Delta\rho_m^2$ may be introduced:

$$\Delta\rho_m^2 = \varphi\left(1 - \varphi\right)\left(\rho_1 - \rho_2\right)^2, \tag{99}$$

where ρ_1 and ρ_2 are electron densities of various microregions of incomplete phase separation, and φ is a volume fraction of one of the phases. The comparison of $\Delta\rho^{2''}$ and $\Delta\rho_m^2$ allows the verification of the suggested microheterogeneous structure.

Together with this value, the microheterogeneity is characterized also by the specific inner surface S/V that can be found from the equation:

$$\frac{S}{V} = \frac{8\pi\varphi(1-\varphi)u^3}{\int\limits_0^\infty \tilde{J}(u)u\,du}\tilde{J}(u)\,, \tag{100}$$

where $\tilde{J}(u)$ is the intensity of scattering, $u = 2\sin\theta/\lambda$ is the scattering vector, θ is the scattering angle, and λ is the wavelength.

The average length of the heterogeneity regions is given as Eq. 101:

$$l_C = \frac{1}{\pi}\frac{\int\limits_0^\infty \tilde{J}(u)}{\int\limits_0^\infty u\tilde{J}(u)\,du}\,, \tag{101}$$

where the integral in the denominator is the mean square of the electron density fluctuation, $<\Delta\eta^2>$. The parameter characterizing the average size of the heterogeneous regions, l_p, is expressed as

$$\frac{S}{4V\varphi(1-\varphi)} = \frac{\lim\limits_{u\to\infty} 2\pi^3 u^4/u}{4\pi\int\limits_0^\infty u^2/(u)\,du} = \frac{1}{l_P}\,. \tag{102}$$

Also, one can estimate the size of microregions with preferential content of one component l_1, the distance between their borders l_2, and their centers $l_1 + l_2$ [131].

A different approach for studying the microphase structure of IPNs based on SAXS and DSC was proposed [132]. This approach uses the mean field theory combined with random phase approximation [51]. The statistical structure factor near the microphase separation for ideal IPNs (where during synthesis no phase separation occurs) is given as

$$S_{MST} = \frac{N_{eff}f(1-f)}{(D(x/6) + x/(x+2)^{-2}) - 2\chi N_{eff}f(1-f)}\,, \tag{103}$$

where $x = s^2 <R^2>$, s is the scattering vector, $s = 2\sin\theta/\lambda$, and $<R^2>$ is the mean-square distance between the chain ends connecting two mutual entanglements, N_{eff} is the number of statistical segments per effective chain, $D(y) = 2y^{-2}(e^{-y} + y - 1)$ is the Debye function, and χ is the thermodynamic interaction parameter. This structural factor has a maximum at some value of x^* that is independent of χ. This maximum corresponds to the scattering vector s^* at scattering maximum. For IPNs the value $x^* = 5.45$ and therefore:

$$s^* = \frac{2.33}{\sqrt{\langle R^2 \rangle}} \, . \tag{104}$$

The characteristic scattering length $l_C = 1/s^*$ and is approximately equal to the mean gyration radius. Authors point out that as distinct from the mean field theories, for real simultaneous IPNs it is necessary to account for phase separation proceeding during IPN formation. This theory considers the spinodal decomposition during IPN formation and introduces the time t_C, after which the decomposition is suppressed. It is assumed that the characteristic reaction time t_C is close to the gelation time t_G (the concept of time during which the separation is possible was developed earlier [133]). The structural factor $S(s)$, reflecting irreversible frustrated fluctuations appearing as a result of competition between phase separation and topological restrictions, may be represented as

$$S(s) = \left[1 - \exp\left(-\gamma \frac{s^2}{S_0} \right) \right] \frac{S_0}{s^2} \, , \tag{105}$$

where

$$S_0 = \frac{1}{\overline{A}s^2 + \overline{m} + \frac{\overline{C}}{s^2}} \, , \tag{106}$$

where A is proportional to the energy gradient, m is the value becoming a structural factor in the limit $s \to 0$, which is proportional to $(T - T_G)$, and C is the coefficient of local attraction between chains. The value γ depends on the monomer diffusion coefficient, component ratio, and kinetics of reaction, characterized by t_C; A, m, and C in this expression are average effective values describing the thermal prehistory of the sample. The correlation length of the microphase can be calculated as

$$l_C = \left(\overline{A} / \overline{C} \right)^{1/4} \, . \tag{107}$$

Since S_0 near $s = s^*$ has the same structure as the structural factor S_{MST}, R_{eff} may be found from Eq. 107.

3.4
Experimental Data on Heterogeneity

Using SAXS, the microphase structure of various IPNs has been investigated. The following IPNs have been studied.

1. PU–styrene–DVB copolymer [130, 134].
2. IPNs based on oligoester acrylates synthesized using anionic polymerization [135].

3. IPNs based on PU and PU ionomer [136].
4. IPNs based on oligoisoprene dihydrazide and epoxy resin [137].
5. PU–PUA [138].

3.4.1
Sequential IPNs Based on Polyurethane and Styrene–Divinylbenzene Copolymer

Figure 22 shows the wide-angle diffractograms of individual networks and IPNs based on them. It can be seen that the characteristic feature of cross-linked PU is the presence of one intensive maximum at 20.2° [130, 134]. The scattering curve of styrene–DVB copolymer has two diffraction maxima at 10° and 19.2°. The scattering curves of IPNs are close to those for PU, if the amount of copolymer is small. However, the shape of these curves differs in the angle range close to 10°. An increased amount of copolymer enhances these differences and later the first maximum of copolymer is seen.

The full similarity of the main diffraction maxima on curves 1 and 2 (Fig. 22) shows that introduction of small amounts of copolymer in a PU

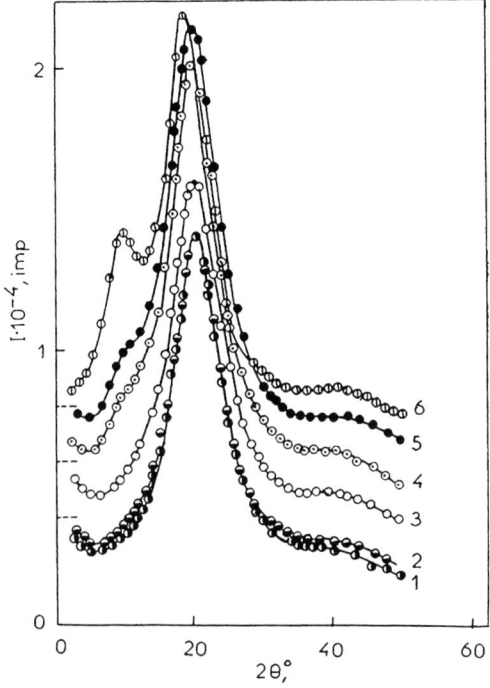

Fig. 22 Wide-angle diffractograms of PU–styrene–DVB copolymer IPNs: (1) pure PU, (6) pure copolymer; IPNs with (2) 3.6, (3) 8.6, (4) 31.4, and (5) 35.4% by mass of copolymer [130]

matrix does not change typical local ordering. For the IPN with 30% of copolymer, the clear appearance of the copolymer maximum shows the existence in the IPN of regions with the copolymer structure.

Figure 23 shows the small-angle scattering intensities (in absolute units) for IPNs and their components. Both for PU and copolymer, the low level of scattering is typical. In the same figure the data for a PU network swollen in monomeric styrene–DVB mixture (53% of this mixture) are presented. For the swollen network the scattering level as compared to initial components is very high. The comparison of the scattering curves (Fig. 23) shows that IPNs fall in between the curves for pure components and those of the swollen network. The increasing amount of the copolymer leads to the growth of the small-angle scattering intensity and to the stepwise change in the shape of the curve. The marked increase of intensity is observed both at very small angles and in the middle part of the curves.

The comparatively small level of scattering for IPNs favors the studies of the heterogeneous structure of IPNs containing even small amounts of the second network. The increasing intensity for IPNs as compared with the additive values for pure components proves the nonhomogeneous distribution of PU and copolymer throughout the volume of the system. The conclusion can

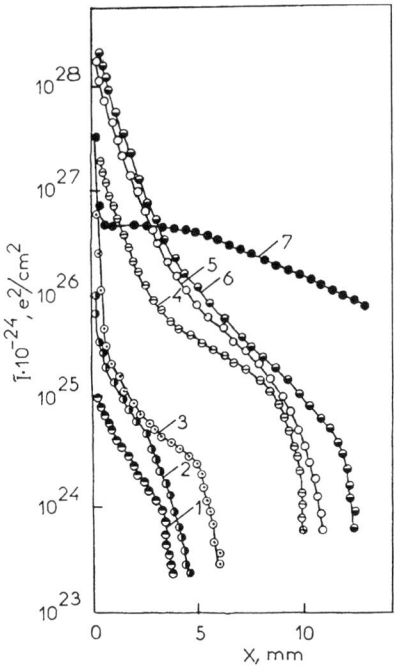

Fig. 23 Intensity of SAXS (in electronic units) for pure copolymer (1), pure PU (2) and IPNs: (3) 3.6, (4) 8.6, (5) 31.4, and (6) 35.4% of copolymer; (7) PU swollen in styrene–DVB mixture (53%) [130]

be drawn that the formation of an IPN leads to the appearance of microregions with different amounts of PU and copolymer. Increasing copolymer amount increases the heterogeneity of the IPN structure.

For the same system the change of SAXS pattern was studied at various stages of styrene–DVB copolymerization. For these systems the parameters of heterogeneous structure calculated from SAXS data are presented in Table 3 [88]. They were determined on the basis of difference curves that were obtained by subtracting additive values of intensities of pure components from the values of small-angle intensities of IPNs.

As we see from Table 3, by transition from small amounts of copolymer to the middle concentrations, the values of l_C sharply decrease at first and then increase. Simultaneously, in the same series the permanent increase of l_1 and decrease of l_2 occurs. The special feature of IPNs with small amounts of copolymer is the sharp difference between values of l_1 and l_2. For the intermediate concentration these values differ only slightly. From Table 3 it follows that the heterogeneity distance, l_C, for IPNs with intermediate composition is much higher than for the same PU network in equilibrium with the monomeric mixture. Increasing the amount of copolymer diminishes the thickness of the transition layer E. Comparison of the data for IPNs of various compositions shows that the increasing amount of the second network (copolymer) results in a significant increase in the degree of segregation, in the diminishing diffuseness of the phase border, and diminishing mixing degree of components.

The experimental data on the heterogeneous structure of the PU network allow us to consider the special features of the heterogeneous structure of IPNs beginning from the stage of their formation. This formation proceeds by the copolymerization of monomers due to microphase separation of PU and copolymer chains. At both small and intermediate concentrations of copolymer in the PU matrix there are microregions of copolymer formed due to the arising immiscibility of network components. As is seen from Table 3, for IPNs with small amounts of copolymer the low segregation degree and rather

Table 3 Parameters of heterogeneous structure of IPNs based on styrene–DVB [88]

Copolymer cont. %	l_C, A	l_1, A	l_2, A	E, A	$\Delta\rho^{2'} \cdot 10^5$, $e^2 \cdot mol^2 \cdot cm^{-6}$	$\Delta\rho^{2''} \cdot 10^5$, $e^2 \cdot mol^2 \cdot cm^{-6}$	α	β	γ
3.6	321	20	270	37	0.307	11.1	0.010	0.972	0.465
8.6	131	26	280	30	5.760	17.9	0.156	0.678	0.515
31.4	170	105	230	23	66.400	73.5	0.657	0.097	0.272
35.4	193	123	225	21	76.700	78.1	0.716	0.018	0.270

high diffusivity of phase borders and high miscibility are typical. Small microregions are the nuclei of the heterogeneous structure. Being about 20 Å in size, they are surrounded by a layer with varying concentration with thickness 30 Å and more. This shows that the main part of the copolymer at small concentrations is located in the transition layer.

For IPNs with amounts of copolymer exceeding 30%, high segregation degree and low diffusivity of the phase border is typical. The comparison of small- and wide-angle diffractograms allows us to conclude that in this case the copolymer forms inclusions in the PU matrix with a size of over a hundred angstroms because of the possibility for copolymer to aggregate into the large particles. At high copolymer concentrations, the conditions for phase separation in the PU-polymerizing mixture system are created, and inclusions are uniformly distributed in the PU matrix. The general conclusions are the following for sequential IPNs at the swelling stage: cross-linked PU is a system of extremely high heterogeneity because of the nonuniform distribution of glycol segments, rigid segments, and solvent (monomeric mixture). In the course of polymerization, depending on the copolymer concentration, various effects are possible. At small copolymer content, the microphase separation remains incomplete. In the intermediate concentration range, due to the high degree of microphase separation, the copolymer inclusions play the role of polymeric filler in the polymer matrix, although the dispersion degree is very high here as compared to usual fillers. In these systems the structural features are reflected in the mechanical properties of such systems [139–141].

3.4.2
Simultaneous IPNs Based on Polyurethane and Poly(urethane acrylate)

IPNs based on PU and PUA are characterized by the fact that the wide-angle diffractograms of both individual networks are very close in the position and the shape of the main maximum, since both networks are purely amorphous systems [134, 138]. Because of this the only criterion of the microphase separation in this case is the SAXS data. Figure 24 shows the intensity of scattering curves for IPNs of various compositions. These curves were obtained, as before, by subtracting the data for pure networks from IPN curves. It is seen that these curves are characterized by the different levels of intensity and shape, depending on the composition. These data show the nonuniformity in the distribution of components in the volume of the system, which is the result of the microphase separation. In the region with a small concentration of one of the networks, the scattering curves have the shape typical for the heterogeneous systems of dilute type. Therefore, at such compositions the correlation in the disposition of heterogeneity regions is absent.

For IPNs with intermediate compositions (40–75% of PUA), increasing scattering in the middle part of the curves is observed, which is characteristic of densely packed systems. However, even in this case, on the diffraction

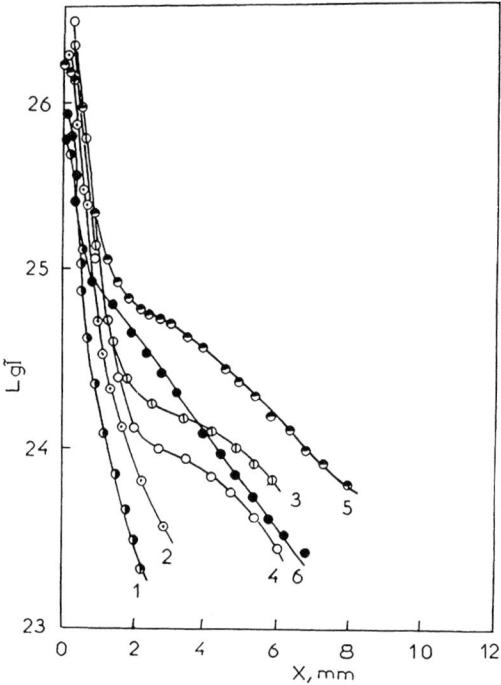

Fig. 24 Dependence of log of absolute intensity SAXS on the scattering angle for PU–PUA IPNs: (1) 5% PUA, (2) 10, (3) 40, (4) 50, (5) 75, and (6) 90% by mass [130]

curves the interference maxima are absent, which could be connected to ordered distribution of microregions of heterogeneous structure. The diffuse character of these curves testifies in favor of various sizes of heterogeneity regions being present.

As in the previous case, the parameters of heterogeneous structure have been calculated (Table 4) [88]. From Table 4 it is seen that the values of l_C in IPNs with concentration of poly(ester acrylate) (PEA) up to 50% are practically unchanged and are close to 500 Å. Simultaneously, transition to IPNs with 75 and 90% of PEA shows the drop of l_C down to 130 Å. This concentration interval (50–75%) is also characterized by the sharp changes in other parameters of heterogeneous structure. Segregation degree sharply increases whereas miscibility of components sharply decreases. Significant changes in heterogeneity parameters in the given composition interval are connected with the phase inversion: with PEA concentration being to 50% of PEA, its inclusions are distributed in the PU matrix, whereas further increase in PEA content leads to the situation where PU becomes a dispersed phase. As the mechanism of phase separation in this case is not known, it is difficult to say if the regions of heterogeneity are almost pure phases or mixed phases that are formed in the course of phase separation.

Table 4 Parameters of heterogeneous structure of IPNs based on PU and PUA [88]

PUA, %	l_C, A	l_p, A	l_1, A	$\Delta\rho^{2''}\cdot10^6$, $e^2\cdot mol^2\cdot cm^{-6}$	α	β	γ	E, A
5	440	320	350	45.2	0.29	0.08	0.70	33
10	475	140	160	50.8	0.31	0.08	0.68	40
40	454	40	70	58.5	0.33	0.07	0.66	20
50	533	80	160	60.2	0.34	0.06	0.65	20
75	123	17	23	128.4	0.75	0.15	0.06	18
90	134	20	25	96.7	0.65	0.12	0.30	19

From the value of l_p and the value of PEA volume fraction, the dimensions of its microregions can be estimated (l_1). The corresponding data are given in Table 4. It should be noted that values of l_1 and l_C are average values over the volume of microregions. However, if in determining l_1 usually the contribution of small microregions is overestimated, in determining l_C there is an overestimation of the contribution of large microregions. Thus, these data allow one to estimate the polydispersity of the heterogeneous structure. From Table 4 it is seen that for 5% of PEA, the values of l_1 are comparable to those of l_C. Therefore, for inclusions of PEA dispersed phase there is observed only a small difference in the sizes of inclusions. At the same time, in the regions of small and intermediate amounts of PU, the values of l_1 and of l_C differ rather significantly, this being the confirmation of the polydispersity of sizes of inclusions. The highest variation in the microregion sizes should be observed in the intermediate range of compositions where the inversion of phases occurs.

Therefore, simultaneous IPNs based on PU and PUA are strongly heterogeneous systems because of the microphase separation of components. The latter is connected to the transition in the course of IPN formation from a miscible mixture of the initial component to a immiscible one, when network fragments begin to appear. It is important that the dimensions of the microphase regions depend on the system composition. In particular, from the data of Table 4 it follows that in the concentration range with a prevailing amount of PEA, the miscibility degree and the interphase layer thickness are smaller, and the segregation degree is higher, as compared to the IPNs with higher content of PU. When PEA is prevailing, the sizes of microregions are lower. These features of microphase structure are connected to the kinetics of polymerization, because the rate of PEA network formation is much higher as compared to PU. Full curing of PEA proceeds in 5 h, whereas for PU the curing time is over 48 h [134]. Therefore, if in the initial oligomer mixture the oligourethane acrylate fraction is no higher than 50%, at the initial polymerization stage the fragments of PEA network are distributed in the liquid matrix of PU components. The sizes of these microregions are 400–500 Å (Table 4). After full curing of PU, the system arises where inclusions of PUA are embedded in the PU matrix. However, if in the initial system the fraction of PUA exceeds 50%, first the complete formation of the PEA network proceeds, and in the PEA matrix the inclusions of fragments of the PU network are formed, their sizes being 10–150 Å. The segregation degree in this case is much higher as compared to IPNs with higher content of PU. For other systems investigated by the same method, similar results were obtained [88, 138–141].

Of a special interest is the process of developing heterogeneity in sequential IPNs where the formation of the second component proceeds in the network which already has its own heterogeneity. Heterogeneous structures of sequential semi-IPNs based on cross-linked PU and PHEMA were studied in [142]. There was established the existence of two hierarchical levels of heterogeneity. The first level relates to the concentration fluctuations inherent to the first early stages of spinodal decomposition and is characterized by evolution of microemulsions of practically pure PU and domains of PHEMA, the latter including the tie chains of PU. The SAXS data have shown various degrees of mixing of two components depending on the IPN composition. The intensive interferential maximum on the scattering curves shows the quasi-periodical distribution of the microregions of phase separation enriched either in hard or soft segments of PU chains.

Authors present the structure of semi-IPNs as consisting of two parts: microregions that have a lamellar structure of initial PU network and domains enriched in a linear component. The dimensions of PU microregions are about 200 Å. These regions are distributed in a heterogeneous matrix that consists of the chains connecting PU microregions and domains enriched in linear polymer.

3.4.3
Sequential IPNs Obtained Using Anionic Polymerization

Of special interest are the IPNs obtained in the course of anionic polymerization [33, 135, 142, 143]. In this case the formation of IPNs proceeds according to the mechanism of "living" polymerization. The following systems were chosen: α,ω-dimethacryl-bis(triethylene glycol phthalate) (MGP) and trioxyethylene α,ω-dimethacrylate (TMA). The second network was a copolymer of styrene and DVB. When IPNs are obtained by anionic polymerization, the second network developing in the matrix of living polymer is formed by an anionic mechanism. A distinctive feature of such networks is the presence of charged active centers and phthalate groups in the main chain. For these systems the structural parameters of heterogeneity have been found using the SAXS method and the theory presented above. The IPNs were obtained under various conditions.

Table 5 shows the structural parameters of the systems under investigation [135]. The data show that the sizes of heterogeneities l_C decrease when

Table 5 Structural parameters of the initial networks and IPNs [135]

System	Co-polymer	MGP	IPN-1-MGP	IPN-2-MGP	TMA	IPN-1-TMA	IPN-2-TMA
l_C, A	1300	1700	1300	1650	840	520	810
$S/V \cdot 10^5$, cm^{-1}	6.84	4.7	6.26	32.12	9.1	16.2	11.7
$\Delta \rho^2 \cdot 10^5$, e$^2 \cdot$mol$^2 \cdot$cm^{-6}	0.148	1.08	2.04	2.56	20.0	16.3	15.6

IPN-1-MGP is formed, while for IPN-2-MGP they remain practically unchanged. The same dependence was observed for IPNs based on TMA, but l_C values in this case were approximately half as large as those in the former case. The specific surface of the heterogeneities for IPN-1-MGP increases a bit, whereas for IPN-2-MGP it increases sharply. For both IPNs based on TMA, an increase in the specific surface is observed. One must pay attention to the increase of the average square fluctuations of electron density by one or two orders of magnitude when passing from copolymer network to MGP or TMA networks.

The analysis of the scattering data shows that all networks are heterogeneous systems of a rather dilute type. The size of the heterogeneous regions in pure networks corresponds, as approximate calculations showed, to hundreds of thousands of oligomeric chain sizes. Creation of such regions can be explained by a microheterogeneous character of polymerization [33]. A twofold increase in $< \Delta \eta^2 >$ in comparison with the network matrix is characteristic of the IPN based on MGP. The above results of the structural investigation of the initial networks and IPNs show that the structure of the latter IPNs may vary considerably, depending on the conditions of preparation.

Next we discuss the interrelation between the structure parameters and the conditions of IPN formation, determined by the successive addition of the components forming the second network and the catalyst. During the formation of IPN-1-MGP a living network matrix was obtained at first, after which styrene and DVB were added. During the initial steps of styrene–DVB polymerization, fragments of PS chains were formed due to transition of electrons from the phthalate rings of the first network. Then the fragments of PS chains were cross-linked by DVB. Coulomb interactions in the system decrease in this case. The ratio of the network matrix to the second matrix was 0.6 : 0.4.

IPN-2-MGP was obtained in a different way, namely, by adding DVB (2 : 1) simultaneously to the living network matrix along with an additional portion of the catalyst. In this case the total concentration of the catalyst for the IPN-1 and IPN-2 remained unchanged. The differences in the synthesis conditions caused the difference in the effective density of the penetrating network, and (in the case of IPN-2) polymerization of the second network was mainly due to introduction of additional quantities of the catalyst, proceeding with simultaneous participation of styrene and DVB molecules. As a result the second network was formed in a different way. Thus, in IPN-1 the cross-linking of the second network had to take place mostly in the regions of the reactive system that had immediate contact with the first network matrix, whereas in IPN-2 the process could proceed within the entire volume all at once.

It was concluded that, because of the heterogeneity of the first network, the distribution of the monomer molecules, which form the second network, is not uniform. Areas with low density of cross-links are places of higher concentration of monomer, while more dense areas contain less monomer. It was established [142, 143] that Coulomb interactions in the growing charged chains play an important role in the formation of anionic networks. For IPN-2, the second network is formed with much stronger intensive Coulomb interactions, as the addition of the catalyst promotes the retention of charges of the phthalate rings of the network matrix. Styrene polymerization proceeds independently, and Coulomb interactions with the charged groups of the network matrix increase. The experimental data show that in both cases, the formation of anionic IPNs raises the inhomogeneity compared to that of the initial network (increase in $< \Delta \eta^2 >$). For IPN-2 the l_C parameter remained unchanged because of strong Coulomb interactions and higher density of the second network, while the specific surface increases sharply (about seven times). This is due to formation in the IPN of smaller areas containing the material of the initial network. The initial heterogeneities area remains preserved.

IPN-1-TMA was obtained with a styrene–DVB ratio of 2.5 : 1, whereas IPN-2-TMA was with a ratio of 7 : 1. The fraction of the second network was about 0.1. In the first case the initial network was first "activated" by the catalyst, and then monomers of the second network were added. IPN-2-TMA was obtained by adding monomers and catalyst to deactivated network matrix. In this case an increase in $< \Delta \eta^2 >$ was not observed. The size of the heterogeneous regions decreases when the relative content of DVB increases (IPN-1). This fact was explained by an absence of Coulomb interactions between the polymerizing chains and the charged fragments of the network matrix, as there are no phthalate rings in the matrix. Therefore, the newly formed network is more likely to be introduced into the network matrix, which results in the smoothing of the difference in electron densities and in decreasing of the size of the heterogeneity regions.

These data show again that the special features of the network formation, even not considering the conditions of phase separation in the system, contribute much to the formation of the second network in sequential IPNs. In the present case, this contribution is influenced by the presence of a great number of charges and by different intensities of the Coulomb interactions between the charged fragments in the network matrix and the growing network. These effects are coupled with the specific growth conditions of the second network in the network matrix due to its considerable microheterogeneity. As a result, IPNs are characterized by different levels of microheterogeneity, by different sizes of heterogeneity regions, and the sizes of their surfaces.

In [144] the structural heterogeneity and segmental relaxation in a series of IPNs of various compositions made of poly(propylene oxide)-based PU and butyl methacrylate–triethylene glycol dimethacrylate copolymer have been studied and compared to the properties of pure constituent networks. The segmental relaxation was investigated in the temperature region around T_g using the laser-interferometric creep rate spectroscopy (CRS) technique. The spectra obtained over the temperature range from 150 to 360 K allow us to characterize in detail the heterogeneity of segmental dynamics within or near the extraordinarily broad glass transition region in the IPN. The systems under investigation do not show distinct microphase segregation and are characterized by a single very broad glass transition. In this work for the first time a number of relaxations were differentiated. The discrete responding of a creep rate spectrum to each of these relations, associated with cooperative, partly cooperative, or noncooperative Kuhn segment movements, has been observed. Combining the CRS data with some data on DSC has allowed the molecular assignments of the multiple creep peaks to be found. The data distinctly show the nanoscale compositional heterogeneity associated with relaxation in the neat networks and segmentally mixed nanodomains. The data also confirm the incomplete compatibility for all IPN compositions.

A special interest lies in the process of developing heterogeneity in sequential IPNs where the formation of the second component proceeds in the matrix network, which already has its own heterogeneity. The heterogeneous structure of semi-IPNs based on cross-linked PU and heterogeneous HEMA was studied in [142] using the SAXS method. The existence of two hierarchical levels of heterogeneity was established. The first is determined by the concentration fluctuations, which are inherent to the early studies of spinodal decomposition, and is characterized by evolution of microregions of practically pure PU and of domains of PHEMA. The latter include some tied chains of PU. The SAXS data show various degrees of mixing of the two components depending on the IPN composition. The intensive interference maxima on the scattering curves show the quasi-periodical distribution of microregions of phase separation enriched in either hard or soft segments of PU chains. The structure of IPNs may be presented as consisting of two parts: microregions that have the lamellar structure of the initial PU network and domains enriched in linear polymer. The dimensions of the PU microregions are about 20 nm. These regions are distributed in a heterogeneous matrix that consists of the chains connecting PU microregions and domains enriched in linear polymer.

3.4.4
Microphase Structure of IPNs Based on PU and Ionomeric PU

In monomeric IPNs the specific interactions between constituent networks contribute strongly to phase separation [145]. However, only limited experimental data exist for such systems. The comparison of wide-angle diffractograms of the individual networks and those of IPNs containing 50% of PU ionomer (copolymer A, see [124, 125]) shows that the PU network has one diffuse peak, whereas ionomers and IPNs do not exhibit such a peak. Small-angle diffractograms of the initial cross-linked polymers and those of the IPNs of various compositions are presented in Fig. 25 [136, 146]. The analysis of these curves indicates the considerable difference between the curves for individual networks and those for IPNs. The scattering curve of the non-ionic PU is notable for the high intensity of its small-angle diffraction and for its broad, pronounced peak. At the same time, for the PU ionomer the scattering level is relatively low and the peak is weak. The angular positions of the interference peaks of both curves are very different. The interplanar distances corresponding to the small-angle peak of the individual networks

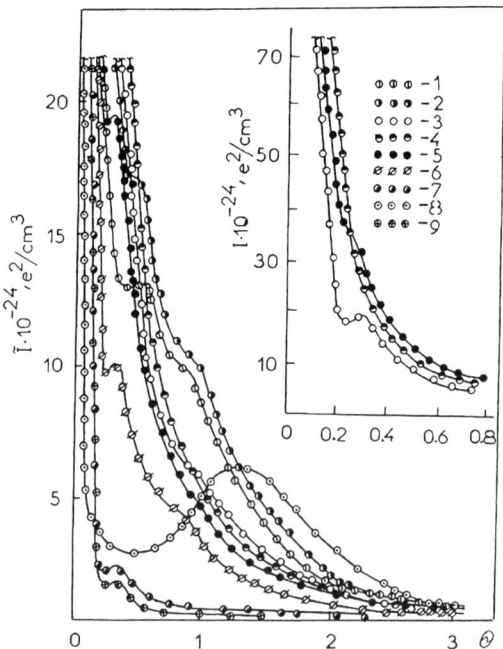

Fig. 25 SAXS diffractograms of the initial network and of IPNs: (1) IPN with 2% of PU ionomer; (2) IPN with 10% of ionomer; (3) IPN with 40% of ionomer; (4) IPN with 50% of ionomer; (5) IPN with 60% of ionomer; (6) IPN with 80% of ionomer; (7) IPN with 98% of ionomer; (8) PU network; (9) PU ionomer network [136]

are correspondingly 7.1 nm for PU and 27.9 nm for PU ionomer. The intensity curves for all individual networks have a common feature—the existence of interference maxima. These curves differ in scattering level and peak positions. Introduction of a small amount of ionomer (2%) into the PU network leads to the substantial "deformation" of the small-angle scattering curve. The pronounced intensity peak typical of PU disappears. In general, the intensity curves become more diffuse, and their features shift to smaller angles. The diffraction curve of IPN shows two weak maxima (Table 6) [136]. Addition of 10% of PU ionomer to PU is accompanied by a general increase of scattering level over the entire small-angle region, but it does not result in a qualitative change of the scattering curve. The IPN with 40% ionomer is distinguished by a very weak diffraction peak at a scattering angle corresponding to an interplanar distance of the order of 10 nm and by an inflection point corresponding to the interplanar distance of 30 nm. IPNs with 50% of ionomer show the opposite effect: the first peak is more pronounced than the second one. Nevertheless, the distribution of interplanar distances remains unchanged and a similar shape of the scattering curve is observed for the IPN with 60% of ionomer. For IPNs with 80% of ionomer, the small-angle diffraction curve drops sharply with increasing angle. The curve for the IPN with 98% of ionomer has only one intensity peak, its position being practically the same as that for the pure ionomer, and at the same time the scattering curve lies much higher than that of the pure ionomeric network. Thus, the main differences in the small-angle scattering curves between IPNs and individual networks are the following: (1) the curves for IPNs are more diffuse and show a considerable shift of diffraction intensity to small angles, and (2) all the curves have two relatively pronounced intensity peaks. The mean-square electron density fluctuations are more dependent on the IPN composition. An increase of nonionic PU content leads to increasing of the mean-square electron density fluctuations (Table 6).

As we see, the wide-angle diffractograms of both the nonionic PU network and the PU ionomer indicate their amorphous character. The fine struc-

Table 6 Parameters of heterogeneous structure of ionomeric IPNs [136]

% of ionomer	d_1, nm	d_2, nm	$\Delta \rho^{2'} \cdot 10^5$, $e^2 \cdot mol^2 \cdot cm^{-6}$
2	18.6	9.7	0.946
10	20.3	9.9	0.875
40	30.0	10.1	0.791
50	31.0	–	0.706
60	33.1	–	0.582
80	30.8	11.4	0.306
98	29.7	–	0.036

ture is determined only by the very short-range ordering of molecular chain fragments. However, the very high scattering intensity and pronounced small-angle diffraction maximum indicate the absence of pure one-phase state in the nonionic and ionic PU, showing that the individual networks may be considered as systems with relatively pronounced microheterogeneity, which is characterized by a definite segregation degree that is equal to about 8% for PU and about 1% for ionomeric PU. Thus, the data obtained show that both individual networks deviate greatly from an ideal two-phase structure. In ionomer, this deviation shows only at the initial stages of phase separation.

The wide-angle X-ray data show that IPNs are also amorphous systems, but at the same time the individual heterogeneous structure of networks is essentially modified due to IPN formation. There are pronounced qualitative differences between the diffractograms for IPNs and those for individual networks. The IPN diffractograms are more diffuse and often have more than one maximum. Analysis of the interference peaks appearing on the small-angle scattering curves for systems containing 2–10% of ionomer allow the following conclusion to be drawn. The first peak shifts to the large-angle scattering region, as compared to that on the ionomer intensity curve, which corresponds to the smallest interplanar distance. If this maximum was the result of the supramolecular ionomer reorganization, its intensity would be negligibly small. In reality, the first peak is comparable with the peak for pure ionomeric PU. It seems plausible to suggest that the introduction of comparatively small quantities of ionomer leads to the formation of heterogeneous microregions. The periodicity of microregion distribution is determined by microphase separation. The first peak corresponds to the heterogeneous structure periodicity, most probably connected with the appearance of microregions with different compositions. The systems with predominant (80% and more) ionomer content show the first maximum near the interference peak in the ionomer diffractograms. This suggests that in this range of compositions, the IPNs tend to show a structural periodicity similar to that of the ionomeric PU. The second peak, as in the case of small ionomer content, should be regarded as a manifestation of the periodicity at the level of nonionic PU blocks of various chemical natures. For intermediate IPN compositions, the peaks on the small-angle scattering curves are less pronounced. This might be the result of considerable periodicity distortions, either at the level of PU and PU ionomer distribution, or at the distribution level of soft and hard blocks in the IPNs.

Additional information may be obtained by comparing the experimental data on $\Delta\rho^{2'}$ and calculated values of $\Delta\rho_c^2$ for various models of the IPN structure. The best description of the heterogeneous structure of IPNs may be achieved on the assumption that the degree of segregation in microregions of PU network and PU ionomers is the same as the degree of segregation characteristic of individual networks, despite the fact that the microregions in real IPNs are more diffuse than those given by the model.

To obtain a more sophisticated model of the heterogeneous structure, it should be taken into account that no complete separation of PU and PU ionomer occurs in these systems on the microlevel. It is most probable that the high degree of nonionic PU and PU ionomer mixing is typical of the IPNs, when phase separation stops at some intermediate stage. This "freezing" of phase separation in IPNs, just as in the case of the individual networks, is determined by a restricted mobility of molecular chains during network formation. This leads to the formation of interconnected structures in the bulk of the polymer, which is a sign of the initial decomposition stages in the region of the unstable states. These IPNs, being practically four-component systems (made of two flexible and two rigid blocks), are much more complex with regard to their phase state than the individual networks. In the course of IPN formation, both the chemically cross-linked PU network and the network of PU ionomer, where cross-links are formed by ionic bonds, undergo phase separation. Since these systems are unstable at this stage, phase separation occurs through a selective growth of composition fluctuations. As during curing the diffusion coefficient diminishes to zero at a certain stage of transformation, phase separation is stopping at some stage that is far from being complete. In such a way the nonequilibrium phase separation proceeds and the nonequilibrium phase separation microregions appear. At some later stage of phase separation, the coalescence of the microregions proceeds, which leads to a distortion of the microregion distribution periodicity typical of the initial stages of phase decomposition. Evidently, the conditions existing in the range of intermediate compositions are most favorable for this coalescence. Thus, IPN formation in this case leads to considerable distortions of the inherent heterogeneous structure in nonionic PU following the addition of the ionomer. The data considered prove once more that the formation of the nonequilibrium diffuse microregions and the incompleteness of the phase separation are the fundamental characteristics of IPNs.

3.4.5
Reaction Conditions and Microphase Structure

The interrelation between the kinetics of IPN formation and the microphase separation is very important to estimate the effect of reaction conditions on the microphase structure of IPNs. An attempt to establish such a relation was done for simultaneous and sequential semi-IPNs based on PU and PBMA [147]. Simultaneous and sequential IPNs of the same composition have been compared. The two reactions were carried out under different conditions. The composition and the structural characteristics of these semi-IPNs are given in Table 7. In the cited work, the SAXS invariant Q, which characterizes the relative degree of the contrast of the heterogeneous structure, the specific inner surface S/V, which is a measure of the heterogeneity size (the

Table 7 Composition and structural characteristics of semi-IPNs [147]

Sample	PU/ PBMA	$[I]·10^2$, mol·l^{-1}	$[kt]·10^4$, mol·l^{-1}	d, nm	Q	E, nm	S/V, cm^{-1}
1	75 : 25	0.74	1.4	6.0	39	0.6	0.0045
2	75 : 25	5.40	1.4	6.8	36	0.5	0.0040
3	75 : 25	10.80	1.4	8.7	37	0.8	0.0035
4	85 : 15	5.40	1.4	6.3	41	0.6	0.0040
5	65 : 35	5.40	1.4	8.8	34	0.8	0.0040
6	75 : 25	5.40	7.0	6.3	31	0.5	0.0040
7	75 : 25	5.40	0.7	7.3	47	0.6	0.0050
8	75 : 25	5.40	1.4	6.5	35	0.5	0.0040

greater the size, the larger S/V), and the thickness of the interphase E, were determined. The conditions of semi-IPN formation can be divided into two groups: simultaneous polymerization (samples 1–6) and sequential polymerization (samples 7, 8). As an initiator, 2,2-azobisisobutyronitrile (AIBN) was used, whereas dibutyl tin dilaurate (DBTDL) took the role of catalyst for PU formation.

When analyzing scattering data, it is necessary to keep in mind that in PU, formed by flexible and rigid blocks, segregation into two microphase regions also takes place. Therefore, the microphase separation in semi-IPNs is superimposed on the microphase structure of the PU network. Figure 26 presents SAXS scattering patterns of some semi-IPNs, listed in Table 7. The comparison of the scattering curves of the initial PU (curve 9) with those for

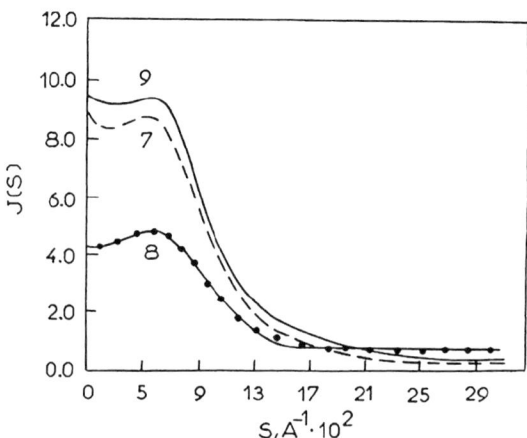

Fig. 26 SAXS scattering patterns of PU network (curve 9) and semi-IPNs (numbers correspond to numbering in Table 7) [147]

a sequential IPN (sample 8) shows that scattering intensity is retained, but the position of the peak is shifted toward the region of smaller angles, corresponding to larger macrolattice spacing (see Table 7). At the same time, the maximum is wider and more diffuse. All these changes are accompanied by a decrease in contrast of the heterogeneous structure and in the thickness of the transition layer, as well as by an increase of the specific inner surface. There is no basic difference between the scattering curve of the pure PU network and that of the semi-IPN. This testifies to the fact that the separation of components during formation of the semi-IPN occurred on a micrometer scale beyond the resolution capabilities of the SAXS method (1 to 100 nm). The swelling by BMA and its subsequent polymerization reduced the periodicity of the PU microphase structure. Some differences in the conditions of synthesis of sequential IPNs are reflected in their structural parameters (Table 7). For simultaneous IPNs these parameters are strongly dependent on the kinetic conditions of reaction governed by different amounts of the initiator and the catalyst. Figure 27 illustrates the effect of PU/PBMA ratio on SAXS pattern. In comparison with the PU network (curve 9), the semi-IPNs have reduced scattering intensity. The period d of the microphase structure is smaller in samples containing 25 and 15 mass % of PBMA and is larger in samples with 35 mass % of PBMA. In general, the spacing of PU increases with PBMA content, but this increase is far from uniform: 0.5 nm for an increase from 15 to 25% of PBMA against 2.0 nm for a jump from 25 to 35 mass % of PBMA. Here, again, we observe a nonlinear behavior. The data for both types of IPNs show that the conditions of phase separation and the microphase structure depend not only on the method of synthesis and reaction kinetics, but also on the PU structure (where microphase separation of rigid and flexible blocks also takes place). Kinetic parameters are very important for the formation of the microphase structure.

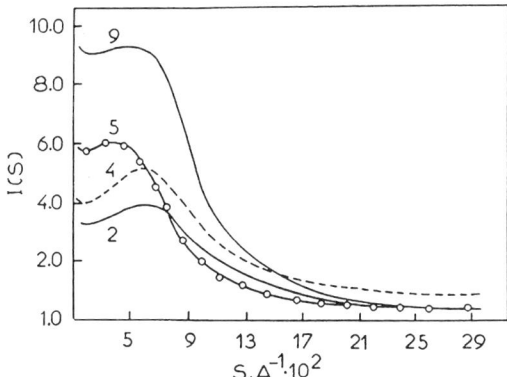

Fig. 27 SAXS patterns of PU network (curve 9) and semi-IPNs (numbers correspond to numbering in Table 7) [147]

Using another approach for estimating the microphase structure, the simultaneous IPNs based on poly(carbonate urethane) (PCU)/PMMA were studied [132]. The authors take for granted that the microphase separation proceeds according to the spinodal mechanism, although they give no evidence. The scattering intensities $I(s)s^2$ were measured. These intensities show growth for small scattering vectors, indicating the appearance of microphase structure with characteristic scattering dimension l_C that diminishes from 293 to 16 nm when the amount of PCU in the IPN increases from 30 to 80%. Correspondingly, R_{eff} diminishes from 682 to 37 nm. The value of l_C characterizes the dimension of the spinodal structure fixed at the gel point, while R_{eff} is a measure of the distance between two mutual entanglements preventing phase separation. The size values for the microphase regions and the Porod analysis of the scattering data reveal the existence of sharp interfaces. These results are in good agreement with the data obtained by Monte Carlo simulations for the IPNs of the same class. The main conclusion made by the authors is that the microphases are the result of microphase separation (proceeding according to the spinodal mechanism) and of the frustration of fluctuations due to an increasing number of topological restrictions in the course of IPN formation. The transition of the microphase separation from a disordered state to the ordered one is the result of the repulsion between both components and the elastic attraction force, arising due to microscopic deformations of entangled chains.

In [148] the SAXS method was applied for the estimation of the structure of an interphase in PU–PS IPNs, in which interpenetrating grafting via HEMA residues has been carried out (1, 2.5, 10% by mass). It was found that with increasing interpenetrating grafting the average size of domains becomes smaller. The data of temperature-modulated differential scanning calorimetry (TMDSC) and dynamic mechanical thermal analysis (DMTA) show that the degree of component mixing increased, and at 10% by mass of HEMA the $\tan \delta$ vs temperature plot showed a single peak. The latter implies that the morphology of this grafted IPN is homogeneous. For three IPNs the correlation function was calculated from Fourier transformation of the scattering intensity for the undefined morphological structure, and the correlation function is given by the following empirical equation:

$$y(r) = \exp\left(-\frac{r}{A_c}\right). \tag{108}$$

The value of A_c represents the correlation distance defined as the size of the heterogeneity in the system. From the dependence of the scattered intensity $I(s)^{-1/2}$ vs s^2, where $s = 2/\lambda \sin \theta$ (λ is the scattering wavelength and θ is the scattering angle) values of A_c were determined. It was found that although the internetwork grafting increases, the correlation length changes very little.

3.5
Characterization of IPN Structure via Small-Angle Neutron Scattering

Besides SAXS, the small-angle scattering of neutrons may be used for studying the structures of amorphous polymer networks. SANS is a very important tool for evaluating polymer chain conformation, structure, and morphology. The first comprehensive literature review on this subject belongs to Sperling [149]. The SANS method permits the examination of bulk materials, especially when a portion of material is deuterium based, rather than hydrogen based. In general, the materials analyzed by the SANS technique are based on hydrogenated polymer matrix, in which a certain amount of isotopically labeled (deuterated) polymer molecules are dispersed. This method takes advantage of the contrast between hydrogenated and deuterated molecules due to the difference in the coherent scattering lengths of deuterium and hydrogen. At the same time, this is a disadvantage of the method because it needs specially synthesized (labeled) polymers. The SANS technique, which may be used to measure the domain size and the chain dimensions within the domain, was successfully applied for studying polymer blends, heterogeneity in polymer crystals etc.

The way to interpret the experimental data and the details of the method were reported in [150–153]. It should be noted that the theoretical basis for SANS is similar to the theory developed for SAXS. All analytical methods developed for X-ray scattering can be adapted to the case of SANS. Below we follow the presentation of the SANS method given by McGarey [150] who used, in his turn, some theoretical approaches from [151–154].

The neutron ray with the wave vector K_0 is scattered through an angle of θ into a solid angle $\Delta\Omega$. The scattered ray has a wave vector K_1 and the position of the scattering center is described by the vector $\vec{R}$. In elastic scattering experiments the absolute values of K_0 and K_1 are both equal to $2\pi/\lambda$. The scattering vector $\vec{Q}$ can be defined as

$$\vec{Q} = 4\pi \sin(\theta/2)\lambda . \tag{109}$$

The coherent intensity in SANS experiments is described by the cross section $d\sigma/dQ$, which is the probability density $d\Sigma/d\Omega = d\Sigma_{coh} + \Sigma_{incoh}/4\pi$ that a neutron will be scattered in a solid angle $\Delta\Omega$:

$$d\sigma/d\Omega = (|b^2| + |b|^2) + |b|| \Sigma \exp(i\vec{Q}\vec{R})|^2 . \tag{110}$$

This equation may be rewritten in terms of the macroscopic scattering cross section:

$$d\Sigma/d\Omega = d\Sigma_{coh} + \Sigma_{incoh}/4\pi , \tag{111}$$

where Σ_{coh} and Σ_{incoh} relate to coherent and incoherent values of Σ, which is the product of the nuclear scattering cross section σ and the nuclear density of the sample.

In SANS studies of polymers, monomer segments are considered to be the basic scattering units and the scattering may be described in terms of the scattering length of the monomer unit:

$$b = \sum_{1}^{N} n_i b_i , \tag{112}$$

where b_i is the scattering length of center i, n_i is the number of atoms of the i-th center in the monomer, and N is the total number of isotopes in the monomer unit. An important role is played by the value of ρ_B, a scattering length density:

$$\rho_B = b/V_M , \tag{113}$$

where V_M is the volume of the monomeric unit. Using the concept of scattering length density, it is possible to define the scattering contrast factor K_f of a two-component system as

$$K_f^2 = \left(\rho_{B_1} - \rho_{B_2} \right)^2 . \tag{114}$$

This factor is analogous to the electron density contrast in SAXS theory. For the two-phase model it consists of discrete particles embedded in a continuous matrix. If there are N_p particles in volume V_p, then taking into account the interparticle interference, characterized by an interference factor $S(\vec{Q})$, the following equation can be obtained:

$$d\sigma/d\Omega = \left(V_p^2 N_p/N \right) K_f^2 P(\vec{Q}) S(\vec{Q}) . \tag{115}$$

To solve this equation, some assumptions regarding the geometry and the physical characteristics of the particles should be made.

Another quantity can be used for the analysis of the SANS data, namely, the correlation function $\gamma(r)$ [155], which describes the distribution of fluctuations in the scattering power throughout the sample:

$$\langle n^2 \rangle \gamma(\vec{r}) = \langle n_A n_B \rangle , \tag{116}$$

where n_A and n_B are the local fluctuations in scattering power (scattering length density) at points A and B, separated by a vector $\vec{r}$ and $< n_A n_B >$ is the average of all possible products as $\vec{r}$ varies from zero to infinity, $< n^2 >$ being the average square deviation. The scattering intensity in this case is presented as

$$d\Sigma/d\sigma = 4\pi C \langle n^2 \rangle \int \gamma \left[\vec{r} \left(\sin \vec{Q}\vec{r}/\vec{Q}\vec{r} \right) \right] \vec{r}^2 \, d\vec{r} . \tag{117}$$

McGarey [150] has used the theoretical SANS approaches for studying sequential IPNs based on poly(dimethylsiloxane) (PDMS) (host) and styrene cross-linked by DVB (guest). PDMS had a certain amount of end vinyl groups for cross-linking. PDMS of various MM and with different cross-link densities were used. A typical normalized SANS curve, obtained from IPNs after background removal, is shown in Fig. 28. The major feature of the scattering curves is the smooth attenuation in scattering intensity with increasing scattering vector. This curve shows the lack of long-range ordering in these IPNs. The Porod analysis [131] of the curves has been done and Porod radii R_p were determined. It was established that the Porod law was valid for the systems under investigation. According to this law, the scattered intensity can be approximated by:

$$I(\vec{Q}) = K2\pi \left(A_p/V_p^2 \right) Q^{-4}, \qquad (118)$$

where K is a constant and V_p and A_p are the volume and the surface area of the scattering particles, respectively. It was found that the Porod radii R_p of scattering particles, found from the expression:

$$Q^4 I(Q) = \frac{2\pi t T K_f \varphi_p}{(1 - T_w) R_p}, \qquad (119)$$

(t, sample thickness; T, sample transmission; T_w, water calibrant transmission; φ_p, volume fraction of particles) change with IPN composition for various molecular weights of PDMS. As the PDMS content increases to 30%, a decrease in the measured radius is apparent. Obviously, this decrease in the domain size must have been accompanied by an increase in the number of PDMS zones. After passing through some minimum value, the Porod radius

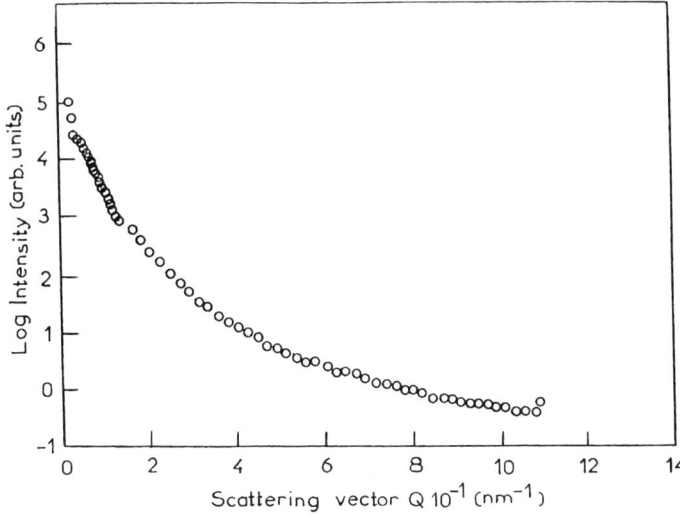

Fig. 28 Typical SANS profile for PDMS/PDS IPN after normalization and background removal [150]

rapidly increases as the PDMS content is increased and at 60% or more of PDMS, the Porod radii exceed 50 nm. The marked increase in Porod radii between 50 and 60% PDMS correlates with the onset of distinct yielding and stress whitening of the IPNs on deformation, both of which can be associated with the guest polymer connectivity. It is possible that the rapid increase in PDMS zone size is due to the fact that the PDMS zones become continuous, invalidating Porod analysis. It was also found that as the host network became more cross-linked, the measured domain dimensions for the guest polymer decreased.

Hosemann [156] developed a method of SANS data analysis, which accounts for polydispersity in the sizes of the scattering particles. This analysis assumes that the distribution of sizes of scattering particles is of a Maxwell type:

$$m\left(R_g\right) dR_g = \left[2/r_0^{n+1} \Gamma\left(n + 1/2\right)\right]\left[R_0^n \exp\left(R_0^2/r_0\right)\right], \tag{120}$$

where $m(R_g)dR_g$ is the proportion of particles with radii of gyration of the scattering particles between R_g and $R_g + dR_g$, Γ is the Gamma function, and r_0 and n are parameters calculated from the scattering profile. The last two values are determined from the plots $I(Q)Q^2$ vs Q. Substituting r_0 and n into Eq. 120 allows the average radius R_{gm} to be found:

$$R_{gm} = r_0 \Gamma(n/2 + 1/2). \tag{121}$$

The results of Hosemann analysis do not show a systematic variation in the average sizes of the PDMS zones as functions of guest polymer content. At the same time, the higher is the cross-link density of the PDMS, the smaller is the size of the guest polymer zones. The more cross-linked PDMS gives rise to domains with greater polydispersity or, alternatively, to domains with greater deviation from a spherical shape. The results of calculations of the domain dimensions are presented in Table 8 [150]. The equivalent sphere radius is determined from the relation $R_g^2 = 3/5R_s^2$.

As is seen from Table 8, there is no correlation between Porod radii and the cross-link density. Over most of the composition range studied, the equivalent sphere radii derived from the Hosemann analysis were significantly larger than the Porod radii. For high PDMS contents, the average domain sizes derived from Hosemann analysis are substantially lower than the Porod radii. The discrepancy between the two methods was ascribed to the invalidity of the models for certain composition ranges.

McGarey has also determined Debye correlation length. He established that the preceding two methods of analysis required the assumption of discrete, spherical scattering particles. The validity of such morphology is questioned and the scattering data may be analyzed by the method developed by Debye and Bueche [155]. This analysis requires the definition of a statistical function describing the distribution of scattering length density in real space. The correlation function of a random distribution of two phases separated by

Table 8 Results of the Porod and the Hosemann analysis of SANS data for PDMS/PDS IPNs [150]

PDMS, crosslink density, mol·m^{-3}	PDMS, volume fraction	Porod radius, R_p, nm	Hosemann average radius, R_g, nm	Equivalent sphere radius, R_s, nm
46.0	0.54	57.1	26.9	34.7
46.0	0.47	19.7	26.1	33.7
46.0	0.37	16.9	27.9	36.0
46.0	0.25	16.3	24.1	31.1
46.0	0.16	20.2	7.4	35.4
66.0	0.64	58.4	15.6	20.1
66.0	0.49	21.7	18.4	23.8
66.0	0.32	15.3	10.9	14.1
66.0	0.16	16.1	19.4	25.0
39.3	0.60	53.5	31.5	40.6
39.3	0.45	33.6	31.8	41.1
39.3	0.38	19.3	24.7	31.9
39.3	0.20	24.3	28.2	36.4

a sharp interface may be approximated by Eq. 108, where A_c is the correlation length and r is a continuously varying position vector. The value A_c may be found from the equation:

$$I\left(Q\right) = 8\pi C < n^2 > A_c^3 / \left(1 + Q^2 A_c^2\right) . \tag{122}$$

Correlation length is related to the degree of heterogeneity of the scattering system via the equation [156]:

$$S_v = 4\varphi_1 \left(1 - \varphi_1\right) / A_c , \tag{123}$$

where S_v is the specific interface surface and φ is the volume fraction of one of the components.

The data converted from the Debye analysis to the equivalent radii agree very well with the results from the Porod analysis over most of the composition range, although at some compositions a discrepancy is observed, connected, probably, with the imperfectness of the models used for calculations.

Generally, the size of the guest polymer zones depends on the IPN composition and varies between 15 and 50 nm. Low correlation was observed between the zone sizes and the cross-linking degree, although there was some evidence for a narrowing of the distribution of the sizes of guest polymer zones when the host polymer cross-linking density was increased. A very important result is that the phases in the IPNs studied were separated by a sharp interface, showing little interaction between phases.

McGarey has also performed a comparison between the experimental guest polymer domain size and its theoretical values, obtained by Donatelli,

Sperling, and Thomas [74]. It was found that the theoretical domain dimensions fail to agree with those obtained from SANS analysis. McGarey suggests that this fact is due to assumptions involved in the derivation of the theory itself. One of the reasons for the discrepancy between theory and experiment is the uncertainty in the interfacial tension, an important parameter of the theory.

Richards [151] considered two IPNs based on cross-linked PDMS and deutero-polystyrene and PDMS–poly(methacrylic acid). In the first case the existence of discrete regions was established, whereas in the second, the formation of a continuous structure of poly(methacrylic acid) in the matrix network was found. The existence of transition layers was discovered for both systems, the thickness of these layers being determined from the scattering data.

In [158, 159] the structures of semi- and full polybutadiene–PS sequential IPNs were studied and the results of determining the equivalent sphere diameter were compared to TEM studies. It was found that the dimensional characteristics obtained by the two methods are in approximate agreement. Surface areas in the range of $150-200 \, \mathrm{m^2 \, g^{-1}}$ indicated true colloidal sizes for the phase domains. Correlation lengths of $35-60$ Å were found for IPNs, $50-100$ and $160-80$ Å for various semi-IPNs. It was also established that these IPNs are characterized by the dual-phase continuity.

Because relatively few experimental SANS data are available for IPNs, presently it is difficult to draw any definite conclusions about the structure of IPNs. As is seen from the data considered, the mechanism of phase separation is not mentioned in any work cited above. Meanwhile, this mechanism should determine if the application of any theory is possible for a given system. One may suggest that the Porod and Hosemann models may be used only for the nucleation and growth mechanism of phase separation, most typical for sequential IPNs. For simultaneous IPNs, where spinodal decomposition, as a rule, is more probable, it seems to be more reliable to determine only the heterogeneity parameters, not the radii of particles, if any. It is also necessary to keep in mind the possible changes of the mechanism of phase separation in the course of reaction.

Unfortunately, there was no attempt to compare the quantitative structure parameters (except segregation degree) with the viscoelastic properties of IPNs. It is more difficult to compare qualitative electron microscopy data with physical properties of the systems.

3.6
Features of IPN Morphology

The morphology of IPNs is determined by the mechanism of phase separation–nucleation and growth or spinodal decomposition. As was discussed above, the application of direct structural methods for studying IPNs is rather restricted due to their amorphous state and low degree of ordering. Because of this, the majority of the investigations of IPN structure were done using electron microscopy. This method, in its turn, has its own restrictions, connected first of all with the difficulties of sample preparation: in many cases the pictures observed are simply an artifact. Many years ago a method was proposed [160] which independently controls the morphology, and which is based on comparison of the morphological picture with the Fraunhoffer diffraction patterns. However, this method has never been applied to real systems. Therefore, electron microscopy gives only a qualitative picture of IPN structure and cannot be used for comparisons of the structure with other properties of IPNs.

Presently, a great deal of work is published concerning the structure of IPNs obtained using electron microscopy. From the preceding discussion it is evident that the morphology of IPNs should be determined by the following factors: (1) the thermodynamic miscibility of two networks, (2) the kinetic conditions of the curing reactions, and (3) the mechanism of phase separation. In principle, three distinct features of IPN structure should be revealed.

The first case is the absence of any structure for miscible and full interpenetrating of different chains. The difficulty consists in the practical impossibility to directly measure the interpenetration degree [161].

The second feature is the structure of the matrix with embedded particles, which is heterogeneous and which shows phase separation in the system. The third feature is the structure of interpenetration phase regions formed in the course of phase separation according to the spinodal mechanism. This structure should be characterized by the presence of a sharp interface between two phases or of diffuse interfacial regions. The appearance of any structure is determined by the phase separation mechanism. The structure is dependent on the concentration ratio of two networks and, depending on this ratio, phase inversion may occur. A typical example is the data obtained for IPNs based on PU and styrene–DVB copolymer [162]. On electron microphotographs there were observed bright and dark regions of irregular shape, which were interpreted as phases of two components. The sizes of the regions of dispersed phase (domains) were found to be in the range $200-500$ Å, depending on the network ratio; inversion of phases was also observed. A sharp heterogeneous structure was observed for full and semi-IPNs based on PU and PMMA. Domains with dimensions of $500-5000$ Å were found. Inversion of phases proceeded at 75% content of PU; however, a clear picture could be observed only if the content of one of the networks was more than 85%. There was no distinct difference in the pictures for full and semi-IPNs. This allowed the conclusion to be drawn that the chain entanglements prevent phase separation and increase the compatibility.

A very interesting observation was made by studying simultaneous IPNs formed by epoxy resin and PBMA [163]. The electron microscopy pictures show distinct two-phase morphology, which allows the dimensions of phases to be determined. It was found that if IPNs were obtained under conditions when epoxy resin was partly cured before the introduction of monomer, the phase separation onset was observed before reaching a gel point, which determined the appearance of the largest particles of the second network. In the course of curing that followed, due to secondary phase separation the smallest particles appeared, leading to bimodal distribution of particles by size. It was assumed that the dimensions of domains depend on the time of phase separation. The smallest particles appear if the phase separation occurred near or after the gel point. Bimodal distribution proved that the reactions really proceeded simultaneously, as the greatest domains appeared before, and the smallest after, the gel point, the domain shape being dependent on the phase separation time. The ratio of the length to the width of domains increased with the amount of cross-linking agent. The simultaneous formation of two incompatible networks from the homogeneous mixture of initial components frustrated phase separation and promoted the seizing of small amounts of one component by large regions of another. Thus, the heterogeneous structure of IPNs was established also by the electron microscopy method [164]. It was also found [165] that the dimensions of domains depend on the amount of initiator and that the morphology is controlled by the time of the onset of phase separation. The domain dimensions are a function of the curing rates and of the time of reaching gel point. The cross-linking density also influences the domain size. The smallest domain sizes are reached by simultaneous curing. The results discussed above have provided a basis for the later investigations of IPN morphology, in spite of the lack of simultaneous investigation of the phase separation mechanism. The main morphological features were formulated by Sperling as follows [165, 166]. Primary phase separation gives the cell structure, which is determined by the onset of incompatibility. In the course of curing, the secondary phase separation proceeds leading to formation of small particles due to high viscosity and diminished molecular mobility. The interpenetration is possible only on the supermolecular level. The largest part of the second phase is disposed in the cells of the first network. Simultaneously, in the matrix network the secondary structure may also develop. Unfortunately, all these concepts had not been confirmed by experiments dealing directly with reaction kinetics and phase separation.

Sperling [16] believes that the results of investigation of many sequential IPNs lead to the following generalizations:

1. Polymer I tends to be continuous in space for all compositions. Mid-range and high concentrations of polymer II tend to have dual-phase continuity.
2. Phase separation probably proceeds through a brief period of nucleation and growth, followed by a region of spinodal decomposition.
3. Domain sizes vary from about 10 nm for highly immiscible systems to about 110 nm for microheterogeneous systems. The domain size decreases with increasing cross-linking density. For many compositions, the effect is about ten times as large for cross-links in network I as for network II.

These principles were developed in virtually all later works. The morphological structures were qualitatively compared with other physical properties. The most important factors are the phase continuity and the phase inversion. A semiempirical approach for predicting phase continuity and phase inversion was developed and studied experimentally [167, 168]. The dual-phase continuity was defined as a region of space where each of two phases maintains some degree of continuity. The model was developed according to which phase (not a polymer!) with lower viscosity or higher volume fraction will form the continuous phase and vice versa. Sperling defines the phase inversion as a transposition of the continuous and the discontinuous phases [16]. The simplest semiempirical rheological relations can be written as:

$$\frac{\nu_1}{\nu_2}\frac{\eta_2}{\eta_1} > 1 \qquad \text{– phase 1 is continuous,}$$

$$\frac{\nu_1}{\nu_2}\frac{\eta_2}{\eta_1} \qquad \text{– dual-phase continuity, either phase inversion,}$$

$$\frac{\nu_1}{\nu_2}\frac{\eta_2}{\eta_1} < 1 \qquad \text{– phase 2 is continuous,}$$

where the subscripts 1 and 2 represent phases 1 and 2, respectively. Experimental data confirm these general relations.

These relations are valid for low shear rates. When either the viscosity ratio or the volume fraction ratio is near 1, dual-phase continuity is encouraged. Both the viscosity and the volume fractions may change with time of curing, so that the morphology will be dependent on the reaction conditions. This correlation was proved for IPNs based on castor oil polyester–urethane and styrene–DVB copolymer. The theoretical phase continuity diagram is shown in Fig. 29. This diagram assumes that the system is stirred when undergoing phase inversion. Yet, because the shear rate is not a part of the model proposed, the shear rate is assumed to be small. The dual-phase continuity was described by Sperling for various IPNs [167–171]. The comparison of the domain diameters in various systems agrees very well with theoretical predictions of domain size.

After the fundamental works dedicated to IPN morphology, many investigations of morphology of various sequential and simultaneous IPNs were published. For example, the simultaneous IPNs prepared from polybutadiene-based PU and styrene–DVB copolymer MMA-co-ethylene glycol dimethacrylate (EGDMA, as the second component) were described [172]. The domain structure was observed, which depends on the ratio of two networks and the distance between the cross-links. The cellular structure and the dual-phase continuity appear in the system as a result of phase separation. In this work the effect of the reaction rates on the morphology was investigated. It was shown that, in general, low simultaneous reaction rates (which is the most typical case for all simultaneous IPNs, see Sect. 5) produce bigger domains. The domain size also depends on the reaction time and the formation rates of each

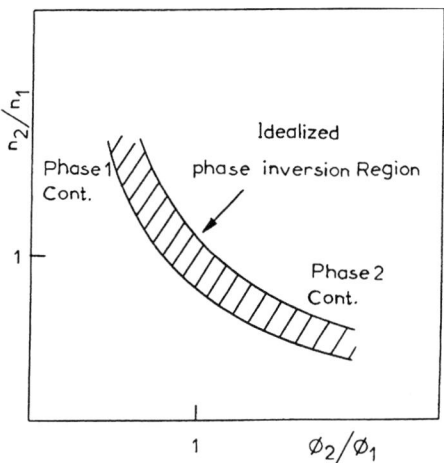

Fig. 29 Theoretical phase continuity diagram. The *hatched area* indicates the approximate range of dual-phase continuity. A phase I (oil-rich) continuous material can be made into a phase II (PS-rich) continuous material by crossing this boundary [15]

network. Only when the relative formation rates of both networks become close to each other are the sizes of the domains moderate and both phases can attain larger cross-linking density. The authors had not observed a strong influence of cross-linking density on the morphology, and after the gel point was reached, no effect was seen.

The review of IPN morphology was carried out by Sperling [16]. He states that for sequential IPNs the presence of cross-linked network I always guarantees that the gelation happens before the phase separation, because the gelation occurs before monomer II is added. If only polymer II is cross-linked, then phase separation precedes the gelation of polymer II. For semi-IPNs the morphology of polymer I may be continuous and of polymer II, discontinuous. For the full sequential IPNs the domains are always finely divided.

Dual-phase continuity, which is defined as the continuity of both polymer I and polymer II domains throughout the macroscopic sample, seems to be better described in terms of spinodal decomposition. In [170] an example of dual-phase continuity was given. PnBA/PS sequential IPNs of various compositions were prepared by UV photopolymerization. PnBA served as network I and PS as network II. Acrylic anhydride (AA) and DVB were used as labile and permanent cross-linkers. After IPN formation, the AA-containing network I was de-cross-linked and solvent extracted. Scanning electron microscopy (SEM) of network II revealed a porous but continuous structure formed by aggregates of fused spherical PS domains. It was shown that network I was continuous, since it could be quantitatively and easily extracted. The major conclusion from this paper relates to the dual-phase continuity of PnBA/PS in sequential IPNs.

The analysis of the morphology development during curing and phase separation was done by Park and Kim [89]. Using SEM, light scattering, DSC, and optical microscopy, the kinetics of synthesis of the polyetherimide (PEI)–epoxy semi-IPN was investigated. SEM was used to study the phase separation mechanism at a given PEI composition. The specimens were taken out at different time intervals during curing, quenched immediately, and then studied. The electron micrographs showed the development of domains during the curing process. The final morphology of samples cured at different temperatures and compositions (5–50 mass % of PEI) was also investigated. The curing of the system with 10 mass % of PEI shows a typical phase separation pattern of nucleation and growth. The coalescence and ripening processes were also observed, in which large particles grow in size at the expense of the smaller ones. When conversion of epoxy reaches 0.4, smaller particles nearly disappear, while the larger particles still grow continuously via coalescence. Another picture was observed for the IPN with 25 mass % of PEI. In this case it was found that the phase separation goes on via a spinodal decomposition mechanism. The viscosity of this system at low epoxy conversion is sufficiently low, thus the interwoven structures coarsen, and spherical particles of the epoxy-rich phase are formed. With increasing epoxy conversion, the particle size increases via coalescence between particles. Finally, the particles touch each other to yield the interconnected nodular structure. For the IPN with 15 mass % of PEI it was found that the initial continuous morphology, which indicates the phase separation via spinodal decomposition, coarsens into dispersed domains, which rapidly grow until they form a macroscale phase-separated morphology. Then, as a secondary phase separation proceeds, epoxy-rich particles and PEI-rich particles are formed in the PEI-rich dispersed phase and in the epoxy-rich continuous phase, respectively, and continue to grow. Finally, the dual-phase morphology, showing epoxy nodular structure, and the "sea-island" morphology are simultaneously formed.

From light-scattering measurements the domain correlation lengths were calculated with the Bragg equation as a function of curing time for various temperatures and compositions. The domain size changes, depending on conditions, between 0.6 and 2.9 µm.

The formation of a dual-phase morphology in PEI–epoxy IPNs is explained in the following way [90, 171]. Two cases were discussed. First, when phase separation is induced by the curing reaction, PEI-modified epoxy shows bimodal UCST behavior (Fig. 30). At composition φ_1 (volume fraction of component 1) phase separation starts at conversion x_1 when the curing temperature is T_1. Because of the spinodal mechanism of decomposition and due to the lower viscosity of the medium, the system will have macroscale phase separation and domains with volume fraction φ_1' and φ_1''. As the reaction proceeds and the conversion x_2 is reached, an abrupt change in the equilibrium composition of the PEI-rich phase occurs from φ_1'' to φ_1'''' in a very short time. This abrupt change is similar to the effect of the two-

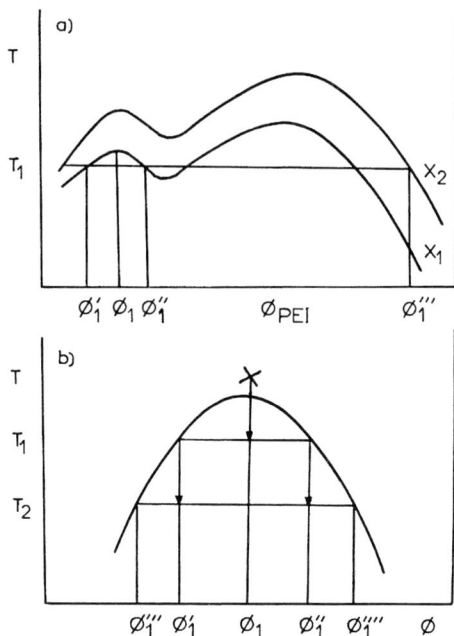

Fig. 30 Schematic illustration of mechanism of dual-phase morphology formation: **a** bimodal UCST; **b** two-step temperature quenching in polymer blend [89]

step temperature quenching in polymer blends having UCST behavior, which results in the dual-phase morphology (Fig. 30b). This means that the first phase-separated PEI-rich composition of φ_1'' now has a higher viscosity and has experienced the jump on the UCST curve, which induces a secondary phase separation within both the domain and the matrix. The cited work is very important because it was clearly shown that the phase domains observed via electron microscopy have the variable composition and are formed by both components of the semi-IPN.

The application of optical microscopy and SEM for studying the kinetics of phase separation in simultaneous semi-IPNs from PU and PS [171] has shown that when phase separation proceeds before the gelation of the medium, PS is dispersed as both large and small nodules in the PU network, and a PU-rich interphase separates the nodules from the matrix. By image analysis it was confirmed that there is a coexistence of two levels of phase separation: a fine segregation of PS in PU and a strong phase separation (PS plus PS in PU). The existence of a shell surrounding the PS nodule was established. At early stages of curing, nodules of PS swollen in styrene coexist with the matrix (PU with styrene). At the end of synthesis the entire monomer has been converted into PS nodules consisting of pure PS. In the cited work [171] the formation of an interphase was observed directly. This interphase consists of a PU-rich zone around the nodule forming a shell. The thickness of the shell has been estimated from micrographs to be around 0.6 μm. The authors analyze the concentration profile of PS for three cases, when nodules are made of pure PS, when the shell contains 39% of PS, and when the matrix contains 67% of PS. The authors believe that the existence of an interphase should involve nearly molecular mixing. It was also found that when the reaction medium remains miscible up to the gel point of PU, only a fine structure exists. Once more, the role of the network formed first in IPNs was confirmed.

The general weakness of all morphological investigations consists of the lack of true knowledge about the mechanism of phase separation both in the cases of sequential and simultaneous IPNs. Because of this the correct formulation of the dependence of morphology on the method of synthesis of IPNs (sequential or simultaneous) and on the phase separation mechanism cannot be done at present. Up to now, it is not clear what mechanism of phase separation is prevailing for sequential and simultaneous IPNs. Since typical spinodal decomposition is observed only in the initial stages of phase separation, it may be possible that at various stages of IPN formation, one mechanism is replaced by the other and that the final morphological picture is the result of more complicated conditions of phase separation, which cannot be described by a single mechanism for all stages of IPN formation. The short review of IPN morphology given in this section shows, as we believe, that morphological investigations alone give no understanding of IPN formation, and that all other methods are needed for correct conclusions.

4
Relaxation Transitions and Viscoelasticity of IPNs

When considering the relaxation behavior in phase-separated IPNs, one has to bear in mind that all the phases should possess their own relaxation characteristics dependent on their composition and structure. This implies the existence of some relaxation transitions, at least two or three of them, one for each phase and for the transition zone. It is also evident that the intensity of relaxation should depend on the cross-linking degree, the free volume, and the segregation degree. Besides, relaxation and viscoelastic properties of polymer networks are dependent on many factors, determining the chemical and physical structure of materials. In phase-separated materials the frequency and temperature variations of the dynamic storage modulus G', dynamic mechanical losses G'', and tangent of mechanical losses $\tan \delta$ are influenced by the morphology of the material and by the nature of the interfaces between the domains of the different phases. The viscoelastic properties of IPNs have been the subject of many investigations. In almost all cases the main goal of such investigations was purely practical—to establish the interdependence between the composition of IPNs and their properties. The estimation of the relaxation behavior of IPNs is a very complicated task and modern viscoelasticity theory cannot explain and describe the properties of multicomponent systems with different morphologies.

The of the most powerful methods in the investigation of heterogeneous polymeric systems is dynamic mechanical spectroscopy (DMS), which enables the estimation of the elastic moduli, mechanical losses, glass transition temperature, relaxation characteristics, etc., and plays a very important role both in the theoretical description of the systems and in their practical application.

Temperature dependencies of loss modulus G'' and $\tan \delta$ usually show sharp maxima at the glass transition temperatures that correspond to the cooperative movement of the segments of polymer chains. Low-temperature maxima are ascribed to the movements of short fragments of the main chain or of the side groups. The equilibrium elastic modulus E_∞ could be an important characteristic of the cross-linking density in IPNs and may be presented as consisting of two parts:

$$E_\infty = E_c + E_e \,, \tag{124}$$

where E_c and E_e are contributions to modulus of chemical cross-links and entanglements, respectively. However, using this equation for estimating the entanglement contributions meets with some difficulties, because the chemical cross-linking in IPNs does not coincide with that calculated from stoichiometry. For semi-IPNs based on cross-linked PU and linear PBMA, it was supposed [174] that $E_c = E_\infty$ (PU) because at all PBMA concentrations in semi-IPNs the glass transition of PU does not change. It was found that under such an assumption, the increasing PBMA content in the IPN increases the contribution of topological entanglements to the elasticity modulus. From these data it follows that the interaction between two evolved phases is realized mainly due to topological entanglements.

When polymer blends and alloys are considered, a problem arises for the application of the principle of temperature–time superposition to the systems consisting of two and more phases. A variant of this superposition has been proposed for such materials [175]. The main feature of the temperature–time reduction in this case is the dependence of the reduction coefficient on the variables—temperature and time. The expression for the reduction coefficient may be obtained via the Taylor expansion of the relaxation function with respect to variable t and T. Phase-separated IPNs relate to thermorheologically complicated materials. In principle, in these systems two different mechanisms of relaxation exist, each of them being characterized by its own temperature coefficient. Because of this, the application of traditional temperature–frequency superposition to IPNs is restricted. However, even in those cases when this approach is not entirely valid, it may be used for approximate calculations. Thus, the application of the temperature–time superposition to heterogeneous polymeric materials shows that the method may be very valuable for prediction of the viscoelastic properties, in spite of the necessity of further developing the theory.

4.1
Application of Mechanical Models

To describe the dynamic mechanical properties of two-phase systems, mechanical models may be used. The most distributed is the Takayanagi model developed for two-phase polymer blends [176]. Two mechanical models describe the elastic properties of the heterogeneous systems, namely, the complex elastic modulus. The corresponding expressions for the moduli are:

$$E_I^* = \left[\frac{\varphi}{\lambda E_A^* + \left(1 - \lambda\right) E_B^*} + \frac{1 - \varphi}{E_B^*} \right]^{-1} \tag{125}$$

$$E_{II}^* = \lambda \left[\frac{\varphi}{E_A^*} + \frac{1 - \varphi}{E_B^*} \right]^{-1} + \left(1 - \lambda\right) E_B^*, \tag{126}$$

where E_A^* and E_B^* are the complex moduli of the two phases, λ and φ correspond to the thickness and length of the specimen occupied by phase, while the product $\lambda\varphi$ is equivalent to the volume fraction of phase A. The higher φ, the closer is the model to ordinary parallel coupling of the elements, and the higher λ, the closer it is to series.

Another so-called isotropic model, proposed by Kraus [177], is characterized by another arrangement of elements in sequential–parallel combination; there are two variants of the isotropic model, which differ in series–parallel and parallel–series coupling of elements.

For the first time the Takayanagi model was used by Frisch and Klempner [178, 179]. Equation 126 was applied to describe the properties of PU–PA IPNs, PA being the matrix the PU is dispersed in. Calculated and experimental data agree rather well for PA/PU ratio 70 : 30. One takes $\lambda = 0.45$ and $\varphi = 0.667$ and for the ratio of networks 50 : 50, $\lambda = 0.88$ and $\varphi = 0.785$. The agreement of such a simple model with the experimental data on IPNs may be considered as evidence of a weak interaction between constituent networks. The data for this system have also shown the appearance of two distinct transitions with slight shifting and broadening of the glass transition. Initially, this fact was considered to be the sign of some level of molecular mixing, but now we understand that this is the result of the formation of two phases consisting of both components, each phase being quasi-equilibrium. Many equations developed for describing the mechanical properties of reinforced polymers [180] may be applied to IPNs. Indeed, it was established that the Kerner equation might rather well describe the mechanical properties of IPNs based on PU and PMMA in the region of a small amount of one of the components [181]. In the intermediate range of compositions, the mechanical properties are better described by the Budiansky equation [182] that has the form:

$$\frac{V_1}{1 + \varepsilon \left(G_1/G - 1\right)} + \frac{V_2}{1 + \varepsilon \left(G_2/G - 1\right)} = 1 , \qquad (127)$$

where V_1 and V_2 are the component volumes, G_1 and G_2 are the shear moduli of the composition and its constituent components, and $\varepsilon = 2(4-5\nu)/15(1 - \nu)$, ν being the Poisson ratio. The applicability of the Budiansky equation reveals once more the two-phase structure of IPNs.

The systems with dual-phase continuity may well be described by the Davies equation [183]:

$$G^{1/5} = v_1 G^{1/5} + v_2 G^{1/5} , \qquad (128)$$

where v_i are the volume fractions of phases 1 and 2.

The review of the first results on the application of the mechanical models to describing IPN properties was done in the well-known monograph by Sperling [2]. For acrylic–urethane IPNs, the Takayanagi parallel model corresponds to the case in which the stiffer component is continuous, while the series model corresponds to the case in which the softer component is continuous. For the IPN mentioned, the experimental data agree best with the parallel model over most of the concentration range.

The Takayanagi model was applied to calculation of the viscoelastic properties of sequential semi-IPNs based on styrene–DVB copolymer/PBMA [88]. Both variants of the model were investigated. The comparison of experimental and calculated data has been done by calculating curves of log G' vs T for various models and by calculating tan δ as a function of temperature for 40% PBMA. The same curves were obtained for other IPN compositions (Figs. 31 and 32). It can be seen that calculated dependencies for the first variant of Takayanagi model agree better with experimental data than those for the second variant. This fact enables one to assume that at a given PBMA concentration, the main part of the copolymer forms some inclusions in the PBMA matrix. Among the possible reasons for the deviation of the calculated values from the experimental ones is the difference in the mechanical properties of the pure constituent components and the same components in the IPN. This is especially true for semi-IPNs with a small amount of PBMA (5% by mass). The heights of the tan δ peaks, corresponding to the relaxation transition of copolymer, depend on the temperature at which the IPNs were formed, whereas the T_g values themselves do not change significantly. Thus, the conclusion may be drawn that the viscoelastic properties of the IPNs depend mainly on the degree of segregation. It was found that the introduction of a small amount of PBMA into the semi-IPN significantly improves its characteristics, which is connected, probably, with the formation of structures typical of spinodal decomposition. In the same work both variants of isotropic models were also used, but the calculated data do not agree with the experimental ones.

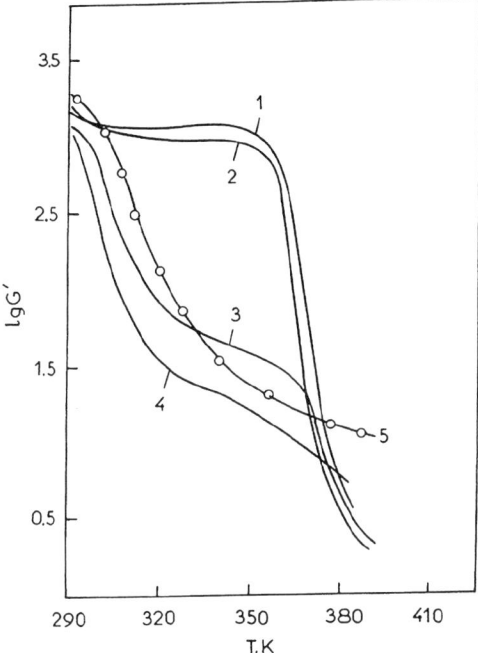

Fig. 31 Temperature dependence of log G' for various models: (2), (4) Takayanagi model; (1), (3) isotropic model. Variant I-1,2; variant II-3,4. Points (curve 5) are experimental values [88]

The Takayanagi model allows calculation of the value of the complex modulus E^*. In [184] the methods of calculating E' and E^* were proposed. For this purpose, using the parameters of the Takayanagi model, the following set of equations was derived:

$$E' = \left(1 - \lambda\right) E'_A + \lambda X / \left(X^2 + Y^2\right),   \tag{129}$$

$$E'' = \left(1 - \lambda\right) E''_A + \lambda Y / \left(X^2 + Y^2\right),   \tag{130}$$

$$X = \left(1 - \varphi\right) E''_A / \left(E'^2_A + E''^2_A\right) + \left[\varphi E'_B / \left(E'^2_B + E''^2_B\right)\right],   \tag{131}$$

$$Y = \left(1 - \varphi\right) E'_A / \left(E'^2_A + E''^2_A\right) + \left[\varphi E'_B / \left(E'^2_B + E''^2_B\right)\right].   \tag{132}$$

Experimentally studied simultaneous IPNs were prepared from PU (based on POPG MM 2000, TDI, and trimethylolpropane (TMP)) and PUA, based on the same components (POPG MM 700) and HEMA. For various component ratios the temperature dependencies of the real and imaginary parts of the complex moduli were determined for pure networks and IPNs. The dependencies of viscoelastic functions calculated from Eqs. 129–132 are presented in Figs. 33 and 34 for individual networks (curves 1 and 2) and IPNs of various com-

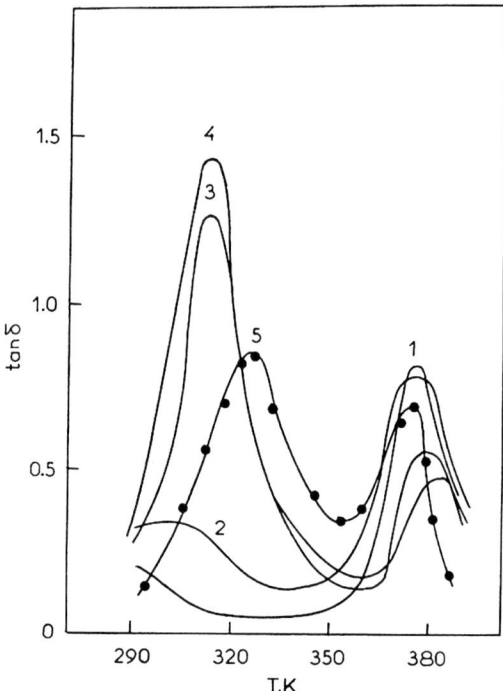

Fig. 32 Temperature dependence of tan δ for the models in Fig. 31 [88]

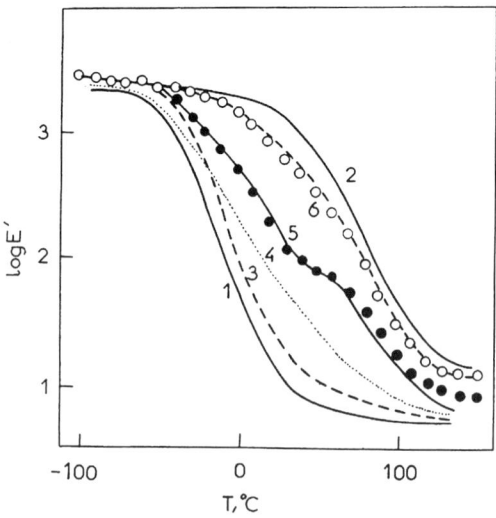

Fig. 33 Dependence of the real part of the complex modulus E' (MPa) on temperature for IPNs of various PU/PUA ratios: (1) pure PU network; (2) pure PUA network; (3) ratio 8 : 2; (4) 6 : 4; (5) 5 : 5; (6) 1 : 9 [184]

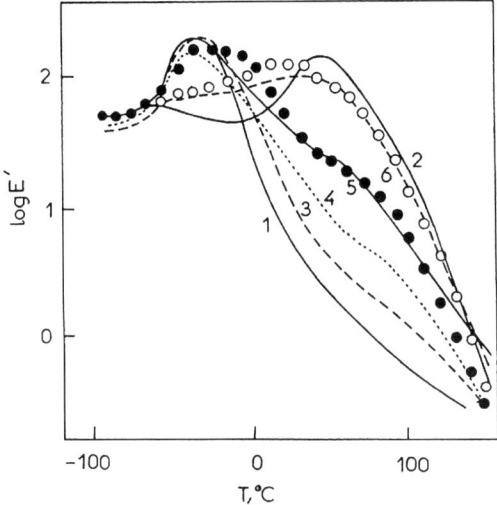

Fig. 34 Dependence of E'' on temperature. Curves as in Fig. 33 [184]

positions (for IPN 4 the volume ratio of PU : PUA is 8 : 2, for IPN 5—6 : 4, IPN 6—5 : 5, and IPN 7—1 : 9).

The temperature dependence of parameter φ (Fig. 35) was determined from Eqs. 129–132, assuming that the continuous phase is a PU-enriched phase. The value of φ was determined as the average of two φ values obtained from the experimentally found values of E' and E''. Applying Eqs. 129–132, a reasonable value $\varphi \leq 1$ could be observed only if one assumes that the PU

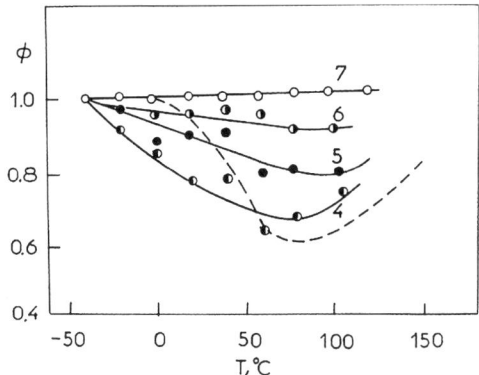

Fig. 35 Dependence of parameter N of the Takayanagi model on temperature for PU/PUA ratios: 1 : 9 (7); 1 : 1 (6); 9 : 1 (5); 8 : 2 (4). *Dotted curve* calculated by Eqs. 124–127 for model of homogeneous mixture ($E^* = 0.5\,(E^*_{PU} + E^*_{PUA})$) with the composition PU/PUA = 1 : 1 assuming that PU is the continuous phase [184]

phase forms a continuous medium up to 90 vol % of the PUA network. Such a conclusion agrees with the data of electron microscopy.

The calculated dependencies of parameter λ on temperature are small and $\varphi \to 1$, which points to additive contributions of the respective phases. The theoretical dependencies of E' and E'' for IPNs 6 and 7 (calculated assuming $\varphi = 1$) are given in detail in Figs. 33 and 34. One can see that good agreements with experiment have been found, particularly for the dependence of E' on T. With samples 4 and 5 no constant λ can be found in order to describe the temperature dependence of the E^* moduli. Thus, the behavior of IPNs increasingly deviates from additivity. It can be supposed that difficulties involved in attempts to adequately describe the mechanical behavior of IPNs by means of a two-phase model may be related to the existence of the interfacial layers, which render the two-phase model inadequate. The results obtained with samples 4 and 5 (Fig. 35) may then be interpreted as showing an increase in the effect of the interphase on the mechanical behavior of IPNs, this effect being much larger for low concentrations of PUA.

The temperature dependence of the parameter φ of the Takayanagi model was also calculated for a hypothetical homogeneous mixture (1 : 1), which would have the value of the modulus $E^* = (E^*_{PU} + E^*_{PUA})/2$. Figure 35 shows that the observed dependencies for IPNs 4 and 5 resemble the assumed shape calculated for a homogeneous mixture. Hence, the temperature dependence of φ for these samples may be due to the reduction of the two-phase character and to a closer resemblance to the behavior of homogeneous systems. Such a conclusion is also supported by additional calculations for the Takayanagi model, for which it was assumed that phase A is the PU whereas phase B is the homogeneous mixture of PU/PUA with the ratio of 8 : 2. Application of Eqs. 129–132 to the E vs T dependence for sample 4 (assuming $\varphi = 1$) led to the temperature-dependent value of volume fraction of the B phase (about 0.70–0.85), which also indicates a considerable homogeneity of this IPN. The results of calculations and their comparison with experiment allowed the conclusion to be drawn that at the volume fraction of PUA > 0.5 the mechanical behavior of IPNs has a pronounced two-phase character, with the effect of the interphase layer being small. At a smaller fraction of PUA, contributions of the interphase are distinct (see below) and experimental results cannot be adequately described by the two-phase model.

The mechanical models were applied to IPNs in acrylic blends and their fibers [185]. The authors have defined a parameter ε to characterize the adhesive capacity between phases, i.e., interpenetration of the two phases:

$$\varepsilon = \sum \left[E'_{cal}(T) - E'_{obs}(T)\right]^2 / 10^{22}, \tag{133}$$

where E'_{cal} and E'_{obs} represent the calculated and the observed modulus, respectively; the summation is taken over the entire temperature range. Using a computer, several pairs of parameters, λ and φ, were calculated. The equivalent models with a minimum value of ε were obtained for model I. In this case, the agreement between calculated values and observed ones is fairly good. The authors believe that the value of ε is related to the compatibility of two network components. The ratio λ/φ can provide some information about the structure of IPNs.

4.2
Viscoelastic Functions and Relaxation Transitions

The phase separation in IPNs determines the appearance of at least two relaxation maxima in the systems, which are determined by the properties of evolved phases. Transition temperatures may be found using all the methods that are usually applied for this purpose, namely, DSC, dilatometry, measurements of viscoelastic functions/temperature dependencies of elastic modulus, and temperature dependence of the tangent of mechanical losses, etc. The appearance of two relaxation transitions was the first sign of the heterogeneous two-phase structure of IPNs and therefore a number of investigations were dedicated to this problem, although when the main features were established, the following investigations could not add anything essential to the understanding of the IPN properties.

As early as the end of the 1960s, Sperling established [186, 187] that for sequential IPNs based on PS and PA, the dependence of the elastic modulus on temperature shows the existence of two transitions indicating the formation of two phases. Transition temperatures are shifted along the temperature scale, which is a sign of incomplete phase separation and of the appearance of an interphase. For some systems it was found [188] that the value of the complex elasticity modulus and its temperature dependence (i.e., the position of relaxation transitions) depend on the way the IPN was synthesized, that is, on which network was obtained first and formed a matrix. The sequential IPNs based on poly(ethyl acrylate) (PEA) and styrene–MMA cross-linked by tetraethylene glycol dimethacrylate were prepared for two cases when the elastomeric network (PEA) or PS network was obtained first. The comparison of the temperature dependence of viscoelastic functions shows rather profound differences, both in values of modulus and position of transitions. These results were explained by various degrees of compatibility of two networks depending on the sequence of their formation.

In principle, similar results were obtained in all subsequent investigations done for various systems [189–191]. As a rule, the temperature transition for the elastomeric component shifts to higher temperatures, whereas the temperature transition for the more rigid component shifts to lower temperatures. Now we can say that these shifts are determined by the formation of two phases with different compositions (initially these shifts were ascribed to the changing compatibility of two components). For semi-IPNs based on PU and PHEMA [191], the segregation degree varies in the limits 0.22–0.35, again showing that phase separation is not completed. Therefore the studied IPNs are two-phase systems with incomplete phase separation and different degrees of phase separation.

The study of viscoelastic properties of sequential IPNs based on PU and styrene–DVB copolymer [192] has shown that increasing the content of the second network affects nonmonotonously the position and the shape of the curves of temperature dependence of the elastic characteristics. This dependence is shown in Figs. 36 and 37. The modulus diminishes only in a narrow

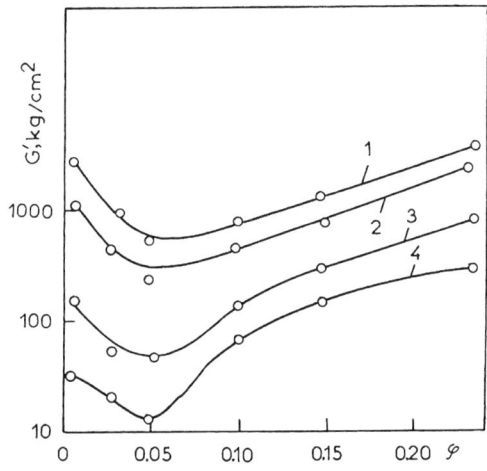

Fig. 36 Dependence of the shear modulus on the volume fraction of the second network at various temperatures: (1) 243, (2) 253, (3) 293 K [192]

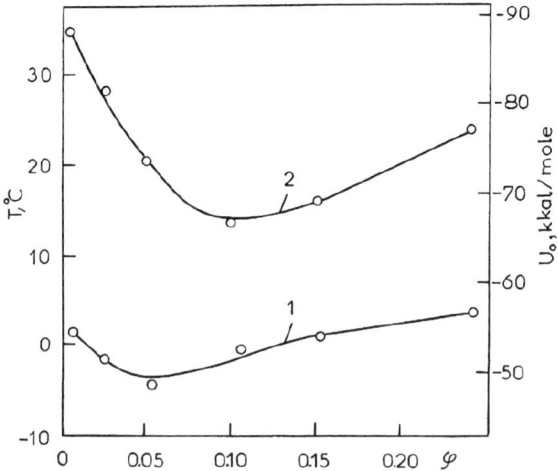

Fig. 37 Dependence of the temperature of maximum losses (1) and activation energy (2) on the volume fraction of the second network [192]

concentration range. The application of the Williams–Landell–Ferry (WLF) theory to these data allowed the temperature dependence of the relaxation times to be calculated and the activation energy of transitions to be found. As is seen from Fig. 37, at rather low content of the second network the activation energy diminishes, but above the volume fraction of 0.1 it increases. The data were explained by the formation of an interphase, the fraction of which depends on the amount of the second network. The formation of an interphase contributes to the viscoelastic properties of the system more, the higher the fraction of interphase. This fraction increases with concentration of the second network. The second component also diminishes the molecular mobility of the chains of the PU network [181]. As a result, the mechanical characteristics of the IPN should depend not only on the content of the second network, but also on the fraction of an interphase. The simultaneous action of two different mechanisms determines the extreme dependence of the IPN properties on the networks ratio. Generally, similar features were discovered for simultaneous and sequential IPNs [165, 193, 194]. The position of the temperature transitions in all cases depends on the reaction time and may be changed rather markedly.

Some specific features are exhibited by IPNs in which one of the networks is formed by linear or cross-linked ionomer [195]. Polymers related to ionomers usually contain some amount of ionogenic groups (usually no more than 10%), as distinct from polyelectrolytes [196]. The way an ionomeric IPN is synthesized affects its viscoelastic properties, as was established for styrene–ethylene–butylene copolymer (copolymer I) and styrene–methacrylic acid–isoprene network (copolymer II) [197, 198]. Copolymer I has two transition temperatures and copolymer II has one, and their shift during chemical mixing is negligible indicating almost full immiscibility of components. The mechanical behavior of such systems is described rather well by the Davies equation (Eq. 128). At the same time, IPNs based on PU and ionomeric PU (the latter was synthesized from 2,4-TDI, 2,2-dimethylethanolamine, and 1,5-dibromopentane) [146] are characterized by comparatively low segregation degree due to some miscibility. Numerous data discussed in [195] show that the distinctions in viscoelastic behavior of ionomeric IPNs are connected with the formation of complexes between polar and ionogenic groups. Generally, the complex formation and very strong trend of such systems to association determine the better conditions for phase separation. The temperature transitions in IPNs were the subject of many investigations, although it is hard to say that new results have added anything essential to the rules established initially.

The relaxation properties of semi-IPNs are characterized by the same general features as for full IPNs [199–202]. The investigation of the temperature dependence of the shear modulus and of the mechanical losses for semi-IPNs based on linear PMMA or polyacrylonitrile and cross-linked PU has shown that the transition temperatures inherent to PU are shifted to higher values when its content is increased. Dielectric measurements have shown that in the temperature interval 173–223 K there appears a weak peak; the β process for PMMA is revealed as a broad peak near 353 K, which moves toward the peak at 403 K corresponding to the α transition for PMMA. For PU two peaks are observed: the first broad one at 183 K, corresponding to the β process, and the second at 248 K, related to the α process, the latter being located 14 K higher as compared with pure PU. From the data obtained the activation energies were calculated for α and β processes in constituent components and in semi-IPNs. It was found that the activation energy for the β process of PU in semi-IPNs is much higher as compared with homopolymer (154 and 100 kJ·mol^{-1}, respectively). This effect was explained by interfacial interaction between components. The authors believed that the smaller are the domain dimensions (determined from the electron microscopy data), the higher is the interfacial area and the more restricted is the molecular mobility at the interface, leading to an increasing activation energy of PU. The ratio NCO : OH in PU and the duration of the time between the onset of PU curing and MMA polymerization also affect the relaxation behavior, the value of M_c in IPNs being also one of the parameters determining the relaxation behavior. The study of dielectric relaxation at low temperatures for IPNs made from epoxy resin and polyacrylate [194] also enabled estimation of the activation energies of the β relaxation process.

It is important that the viscoelastic properties of IPNs depend on the kinetic conditions and on the method of IPN synthesis [202]. This question was thoroughly investigated for semi-IPNs based on cross-linked PU and linear PBMA. The latter polymer was introduced into the network in two different ways:

1. The curing of PU proceeded in the presence of PBMA introduced into the reaction system.
2. Semi-IPNs were formed via simultaneous curing of PU and polymerization of BMA.

The data on the temperature dependence of the elastic modulus and of the mechanical loss tangent are given in Fig. 38. It is seen that semi-IPNs are typical two-phase polymeric systems, exhibiting two relaxation transitions (determined from the positions of loss tangent maximum). The shape of the temperature dependence of elastic modulus is also typical of two-phase systems. At the same time, a substantial shift in T_g of the linear polymer from 343 to 308 K (samples 8–3) and a less marked shift in T_g of the PU network indicate the formation in the system of two phases with dissimilar compositions. The concurrent processes of PU network formation and microphase separation caused by thermodynamic immiscibility of components result in the formation of phases enriched in one of the components.

Some features of the phase structure, which follow from IPN viscoelastic properties, can be pointed out. The continuous medium for samples 2–6 is the phase enriched in PU network, which is evidenced by a plateau on the elastic-

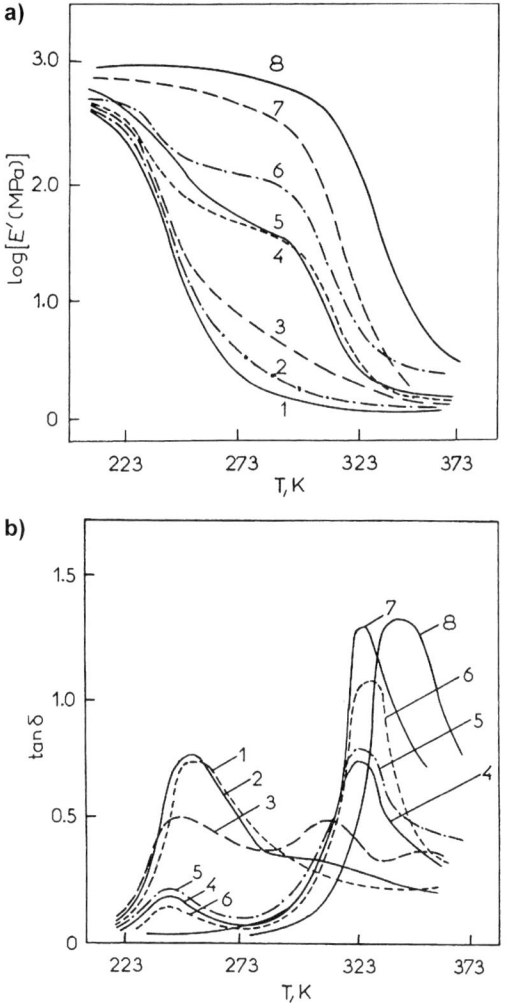

Fig. 38 Temperature dependence of elastic modulus (**a**) and mechanical losses (**b**) for semi-IPN PU/PBMA with ratio of components (mass %): (1) 100 : 0; (2) 91 : 9; (3) 87 : 13; (4) 83 : 17; (5) 77 : 23; (6) 67 : 33; (7) 56 : 44; (8) 0 : 100 [202]

ity curve at temperatures above 343 K. A drastic decrease of the elastic modulus above this temperature for the semi-IPN containing 44 mass % PBMA (curve 7) unambiguously indicates a high PBMA content in the continuous phase. As for many full and semi-IPNs, the inversion of phases enriched in one of the components occurs in the region of intermediate compositions [155].

Since the phase separation proceeds under nonequilibrium conditions, it results in the formation of two continuous phases (dual-phase connectivity) whose compositions are fixed by cross-linking. As was mentioned above, each of the phases consists of both IPN components and can be treated as an independent IPN where mixing of the components at the molecular level through topological entanglements takes place (the "forced" compatibility effect).

4.3
Viscoelasticity and Curing Conditions

The viscoelastic properties are also dependent [202] on the kinetic conditions of curing (Fig. 39). From these data it can be seen that the growth of the PU network formation rate due to an increase of the catalyst mass fraction at a constant temperature 333 K results in the formation of a less cross-linked PU network. The T_g shifts toward lower temperatures (from 261 to 248 K). However, the increase in reaction rate caused by a change in the curing temperature in the series 313, 333, 353 K led to the opposite results. The T_g of the PU network shifts from 251 to 265 K, while the T_g of PBMA remains practically unchanged. The observed changes in the relaxation behavior of the semi-IPNs seem quite logical when considered from the

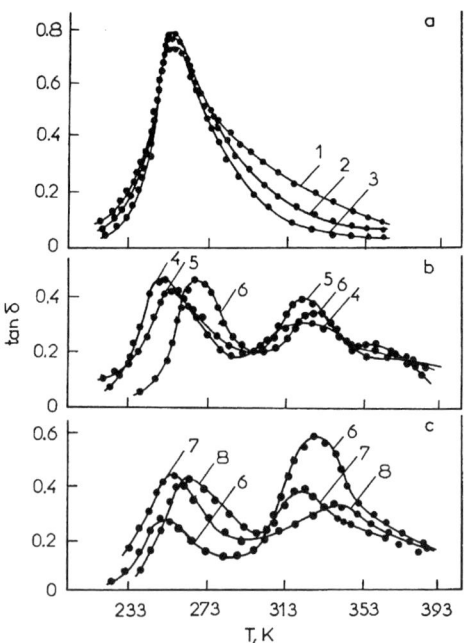

Fig. 39 Temperature dependencies of mechanical losses of PU network (1 – 3) and semi-IPN PU/PBMA (85:15 mass %) (3 – 8) at various temperatures of curing and catalyst concentration (mass fraction *in brackets*). Curves 1 333 K (0.005); (2) 333 K (0.010); (3) 313 K (0.010); (4) 333 K (0.010); (5) 333 K (0.005); (6) 333 K (0.050); (7) 353 K (0.010); (8) 313 K (0.010) [202]

standpoint of concepts of formation of phases enriched in one of the components of a two-phase system. In the above-considered series, the increase in T_g of the PU-enriched phase appears to be associated with the decrease in the reaction system viscosity, which results in the formation of a more regular structure of the PU network. As would be expected, as the curing temperature is lowered to 313 K, the T_g of the PU-enriched phase shifts somewhat to lower values since a more defective structure of the PU network is formed with increasing viscosity of the reaction system. In the viscoelastic properties, a substantial part is played by the method of changing the reaction rate: when this parameter is varied by changing the catalyst concentration, the initial viscosity of the reaction system remains unchanged, whereas changing the curing temperature substantially alters both the reaction rate and the system viscosity, which leads to the formation of a different structure and, as a consequence, to semi-IPNs with different properties.

Now let us consider the same semi-IPNs but produced by simultaneous curing of PU and polymerization of BMA. Comparing two different ways of semi-IPN production, we would have to take into account that PBMA introduced into the reaction system in the first case has a different molecular mass than PBMA formed during simultaneous curing and polymerization. In the cited work this question was not considered, in spite of the possible effect of a difference in molecular masses on the viscoelastic properties. In the case under consideration, both reactions proceed concurrently. Figure 40 shows that with increasing initiator concentration, i.e., with increasing BMA polymerization rate, the temperature of the glass transition of IPNs changes only insignificantly, its shift on the temperature scale being within 2 K. At the same time, a different BMA polymerization rate results in significant changes in viscoelastic properties. A rise in the T_g of PBMA or, more exactly, of the PBMA-enriched phase is observed when the polymerization rate of BMA is increased, deviations toward both higher and lower temperatures (curves 2 and 8) taking place. The T_g of the PU-enriched phase (curves 2, 4, 6, 8) shifts toward higher temperatures from 248 K (initial PU network) to 265 K in semi-IPNs. The observed changes in the viscoelastic behavior evidence an incompleteness of the microphase separation process and a significant influence of the reaction kinetics on the process. The measurements of the elastic moduli for the same systems have shown an increase of E_∞ for samples 2, 4, 6, and 8 with respect to the initial PU network. This fact seems to be associated not only with the contribution of topological entanglements between components, but also with an increase in the intermolecular interaction between them and, possibly, with the increase in the cross-linking degree of PU. This effect differs substantially from the results obtained for the preceding semi-IPN series, where an already macromolecular PBMA was introduced into the system. Evidently, the system viscosity decreases due to the introduction of monomer and gives rise to a more regular network structure, increases intermolecular interaction, and hence raises T_g.

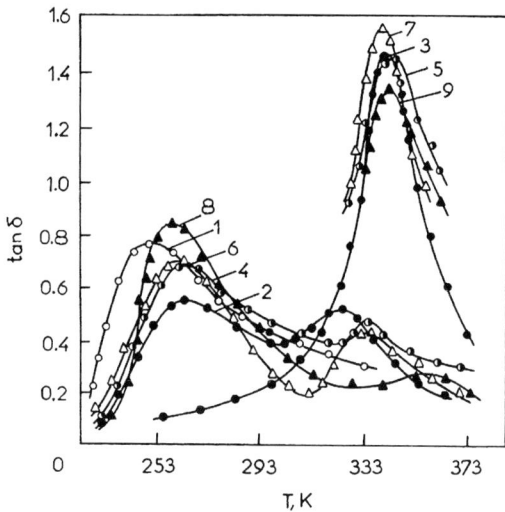

Fig. 40 Temperature dependence of mechanical losses of initial components (PU and PBMA) and semi-IPNs at various concentrations of initiator $[I]$, mol·l^{-1}. (1) PU; (2) PU/PBMA 75 : 25, $[I] = 0.74 \times 10^{-2}$; (3) pure PBMA, $[I] = 0.74 \times 10^{-2}$; (4) PU/PBMA 75 : 25, $[I] = 1.48 \times 10^{-2}$; (5) pure PBMA, $[I] = 1.48 \times 10^{-2}$; (6) PU/PBMA 75 : 25, $[I] = 2.96 \times 10^{-2}$; (7) pure PBMA, $[I] = 2.96 \times 10^{-2}$; (8) PU/PBMA 75 : 25, $[I] = 1.08 \times 10^{-1}$; (9) pure PBMA, $[I] = 1.08 \times 10^{-1}$. Catalyst concentration for PU formation in all cases 1.4×10^{-4} mol·l^{-1} [202]

The shift of T_g of the PBMA-enriched phase toward low temperatures depends on the PBMA formation rate. It seems that the lower is the PBMA formation rate, the more favorable are the conditions for the intermolecular interaction between forming PBMA macromolecules and the PU network being formed, and the lower is the T_g of PBMA in the semi-IPN.

The viscoelastic behavior of the semi-IPNs obtained at the highest PBMA and the lowest PU network formation rates is somewhat unusual. The smallest E_∞ value of the PU-enriched phase indicates a more defective structure of the network. This fact can be interpreted as follows: it is possible that in this case the earlier formed PBMA phase has the form of disperse inclusions, while the PU network is formed in the presence of such a polymeric filler and, as was ascertained elsewhere [180], a more defective network structure develops in such cases.

It seems interesting to compare viscoelastic properties of semi-IPNs obtained with the use of BMA monomer and those made of PBMA. The temperature dependencies of the mechanical losses for these systems are shown in Fig. 41. This figure shows that the method of PBMA introduction (monomer or previously prepared polymer) radically changes the semi-IPN structure and its viscoelastic properties. Of course, in all cases the semi-IPNs are incompatible heterogeneous systems, but significant changes in glass

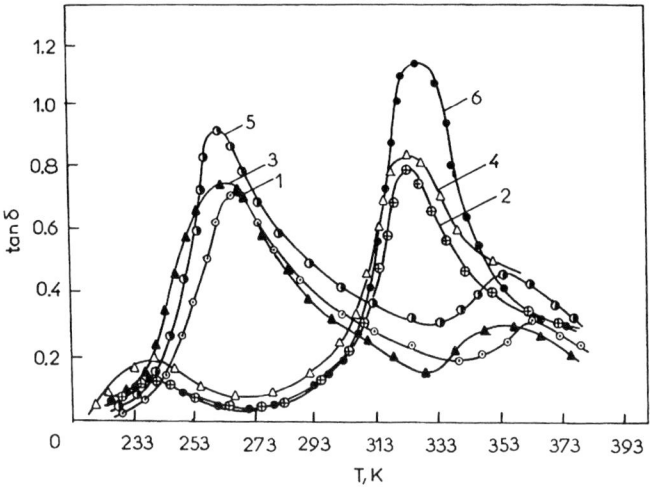

Fig. 41 Temperature dependence of mechanical losses of semi-IPNs at various component ratios: (1) PU/BMA 85 : 15, $[I] = 5.4 \times 10^{-2}$ mol·l^{-1}; (2) PU/PBMA 85 : 15; (3) PU/BMA 75 : 25, $[I] = 5.4 \times 10^{-2}$ mol·l^{-1}; (4) PU/PBMA 75 : 25; (5) PU/BMA 65 : 35, $[I] = 5.4 \times 10^{-2}$ mol·l^{-1}; (6) PU/PBMA 65 : 35 [202]

transition temperatures unambiguously indicate a change in the phase structure of the IPN. The height of the tan δ maximum of a continuous phase always exceeds the corresponding maximum of a phase that is a disperse inclusion. The following can be inferred unambiguously from the temperature dependence of the mechanical loss tangent, presented in Fig. 41. Changing the semi-IPN production method means changing the continuous medium of the system. When PBMA in an amount of 15 mass % is introduced, the continuous medium (matrix) is the PBMA-enriched phase, while the PU-enriched phase is the continuous medium when BMA monomer is used (Fig. 41, curves 1 and 2). Of course, an inversion of phases occurs in both cases, but the concentration range in which this is observed depends on the method of linear polymer introduction into the system. Evidently, the phase inversion in the former case is observed in the region of small PBMA mass fractions (less than 13%), while in the latter case, at a PBMA mass fraction of 35%, the continuous phase is the PU-enriched one.

It was also discovered that the viscoelastic functions depend on the sequence of component curing [202]. In the work cited it was also established that the kinetic conditions affect the segregation degree and, thus, the viscoelastic properties. By increasing the initiator concentration from 0.74×10^{-2} to 2.96×10^{-2} mol·l^{-1} the segregation degree, α, changes from 0.11 to 0.28. In such a way, by varying the kinetic parameters of the component curing and, therefore, the parameters of phase separation, the viscoelastic properties of IPNs may be changed, which is an effective technique for controlling the structure and the properties of hybrid binders of this type.

Knowledge of the general principles that govern the viscoelastic and mechanical properties of IPNs allows us to obtain various polymeric materials on their basis. Because of this, up to now many publications have appeared where these properties are studied [203–208].

4.4
Properties of Thermoplastic Apparent IPNs

Presently, some hybrid polyblends, such as the thermoplastic apparent interpenetrating polymer networks (AIPNs), call for a broader view. In contrast to traditional IPNs, in thermoplastic AIPNs the components are cross-linked by means of physical, instead of chemical, bonds. These physical bonds are glassy domains of block copolymers, ionic clusters in ionomers, or crystalline domains in semicrystalline polymers. The components of thermoplastic AIPNs are capable of forming physical networks and are characterized by mutual penetration of phases. Thermoplastic AIPNs are intermediate between mixtures of linear polymers and true IPNs because they behave like chemically cross-linked polymers at relatively low temperatures, but as thermoplastics at elevated temperature [208]. The blends based on combinations of physically cross-linked polymer and linear polymer, or physically cross-linked polymer and chemically cross-linked (thermoset) polymer, where the physically cross-linked polymer network constitutes the continuous phase and the other component disperses into domains, will also exhibit the properties of thermoplastic compositions.

In the 1990s, the first papers on the synthesis, kinetic peculiarities, and characterization of structure–property relationships for polycyanurate–PU semi-IPNs were published. The idea of these semi-IPNs was to synthesize a new material from widely used linear PU as a thermoplastic elastomer and intensively developing high-performance cyanate ester resins (CERs) as a thermoset. Linear segmented polyurethanes (LPUs) exhibit rubbery characteristics and thermoplasticity that is directly connected with their structure. LPU is characterized by internal heterogeneity specified by phase separation of soft and hard blocks of the polymer chain. CER formulations offer a variety of excellent thermal and other properties (see Sect. 1), which commend them for use in high-performance technology (e.g., high-speed electronic circuitry and aerospace composite matrices) [209, 210]. Some reviews dedicated to thermoplastic PU (TPU) elastomers in IPNs have been published by Fainleib, Grigoryeva, and Sergeeva [210, 211]. These authors contributed very much to the development of the chemistry of TPU IPNs.

Thermoplastic AIPNs of several compositions prepared by mechanical blending in a roll mill of crystallizable polyurethane (CPU) and styrene–acrylic acid random copolymer (S-co-AA) have been investigated using different techniques by Vatalis et al. [212, 213]. The CPU was based on TDI (mixture of 2,4- and 2,6-isomers, molar ratio 65 : 35) and oligomeric butylene adipate glycol (BAG, MM 2000); the molar proportion of reagents was 1.01 : 1.00. The S-*co*-AA was obtained by a bulk radical copolymerization of S and AA; the molar ratio of comonomers in S-*co*-AA was $\sim 72 : 28$. Both individual CPU and S-*co*-AA components are physically cross-linked polymers. CPU is cross-linked by means of hydrogen bonds and microcrystallites of BAG acting as effective cross-linking sites with a degree of crystallinity of about 50%. Won et al. [214] have shown that in S-*co*-AA the significant physical cross-linking occurs by means of strong dimer hydrogen bonding of carboxyl groups.

Some interactions between the components and their effect on the microphase structure of each other in thermoplastic AIPNs are realized, and wide-angle X-ray scattering (WAXS) and SAXS have been reported [212]. The mechanical properties of the thermoplastic AIPNs, density ρ, flow limit σ_f, elasticity modulus E, and tensile strength σ, change nonadditively with composition, with extreme values at small contents (until 5–10%) of CPU or S-*co*-AA in compositions. This behavior confirms that each of the components of the thermoplastic AIPNs affects the microphase structure and properties of the other component. This is only possible by the formation of the interpenetrating structure of the thermoplastic AIPNs [215] by strong physical interactions between the functional groups of CPU and S-*co*-AA. It is known [214] that the acid groups of AA in S-*co*-AA can be considered as anionic groups (COO^-H^+) with a low degree of ionization, which are able to take part in intermolecular physical (electrostatic) interactions with polar groups of the flexible and rigid CPU blocks.

Some secondary relaxation of components of thermoplastic AIPNs have been investigated by using thermally stimulated depolarization current (TSDC) techniques and thermally stimulated conductivity (TSC) measurements [212, 213]. It was found that on addition of S-*co*-AA to CPU, the secondary γ and β CPU peaks (at ~ -140 and $\sim -100\,^\circ C$, respectively) shift slightly to lower temperatures, i.e., the corresponding relaxations become faster, the shifts being more pronounced at low S-*co*-AA contents. The shifts may be brought about in relation to physical interactions between the IPN components and to their partial miscibility. Rizos et al. [216] have shown that as a result of such interactions, changes of local free volume may occur, which affect the relaxation time of secondary relaxations. The same changes of the β relaxation of PU have been found in PU–PS IPNs by Pandit and Nadkarni [217].

The method of mechanical blending of two (or more) polymers in the melt is one of the main methods of producing thermoplastic IPNs [208]. Thermoplastic AIPNs based on the same CPU and S-*co*-AA by mechanical blending of components in a common solvent were investigated by DMTA, DSC, WAXS, SAXS, TSDC, and other techniques by Sergeeva, Kyritsis et al. [218–222]. DMTA measurements have shown that thermoplastic AIPNs can be considered as multiphase systems having at least two amorphous and one crystalline phase and regions of mixed compositions. Their mechanical properties are determined by the heterogeneity of the individual components, as well as by the heterogeneity caused by the thermodynamic immiscibility of the components. The degree of incompatibility is determined, to a large extent, by the ratio of intra- and intermolecular H bonds between the functional groups of CPU and S-co-AA. For thermoplastic AIPNs with CPU content up to 10%, CPU–S-co-AA interactions mainly take place, whereas at higher CPU contents CPU–CPU and S-co-AA–S-co-AA interactions begin to dominate. It is important to note that all the physical cross-linking, i.e., crystallites of CPU and inter- and intramolecular H bonds of CPU and S-co-AA are destroyed at elevated temperature [209].

The results obtained are explained [220] by the formation of the networks of inter- and intramolecular H bonding, which results in formation of double phase continuity. The intermolecular H bonding between the functional groups of the components promotes the improvement of compatibility of the components in the thermoplastic AIPNs. On the other hand, the degree of segregation of the CPU and the S-co-AA microphases simultaneously increases, as AA and BAG move out of the S-co-AA and the CPU phases, respectively, because their functional groups take part in intermolecular H bonding. Intramolecular H bonding promotes a deepening of the microphase separation of the thermoplastic AIPN components.

The thermodynamic state of thermoplastic AIPNs produced by the solution technique has been investigated by Sergeeva et al. [221] to give an insight into their structure as a function of their composition. To estimate the thermodynamic state of the investigated thermoplastic AIPNs, the free energy of mixing the constituent polymers has been calculated by the vapor sorption method. The changes in free energy of mixing (Δg^x) between the CPU and S-co-AA in the thermoplastic AIPNs with the range of composition chosen have been calculated. It was found that the values of the free energy of mixing of components Δg^x are slightly negative ($\Delta g^x = -0.3/-3.6\,\mathrm{J\cdot g^{-1}}$) in the range of small content (up to 10%) of each of the components, and it is positive ($\Delta g^x = +0.5\,\mathrm{J\cdot g^{-1}}$) for thermoplastic AIPNs with CPU content of 50%. Therefore, in thermoplastic AIPNs with a content of CPU or S-co-AA up to 10%, the components are thermodynamically miscible and thus thermoplastic AIPNs are thermodynamically stable. The microphase separation of the components takes place for the thermoplastic AIPNs of middle compositions and, probably, for other compositions where the systems are thermodynamically unstable. One can suppose that in this case the complete phase separation of the components of thermoplastic AIPNs does not occur, this process being prevented by intermolecular physical network formation by the components. Note that in the transition region from the thermodynamically stable to thermodynamically unstable (metastable) state extreme changes of all properties of the thermoplastic AIPNs studied have been observed.

Thermoplastic AIPNs based on the same CPU and (styrene–acrylic acid) ion-containing (K^+) block copolymer (S-b-AA (K)) prepared by mechanical blending of components in a common solvent were investigated using DMTA and DSC techniques by Bartolotta et al. [223]. The local and cooperative molecular mobility of individual CPU, S-b-AA (K), and thermoplastic AIPN of composition CPU/S-b-AA (K) = 80 : 20 (mass %) has been investigated. The experimental results for thermoplastic AIPNs have been compared and contrasted to those obtained in pure components, showing the existence of distinct calorimetric and mechanical transitions which are unambiguously associated to the two components, a clear indication of a multiple-phase heterogeneous structure. These observations reflect a thermodynamic incompatibility of the components, even though a limited miscibility is inferred by small but appreciable variations of the magnitudes of local (γ relaxation) and cooperative (α_a relaxation and melting events) transitions, which deviate significantly from a simple dilution effect. The shift of T_g of the CPU component to higher temperatures and the largest jump of ΔC_p observed in the DSC thermogram of thermoplastic AIPN, as compared to the thermal features characterizing pure CPU, as well as the decreasing crystallinity of the CPU component, are the results of limited miscibility of the components. The growth affinity between CPU and S-b-AA (K) in thermoplastic AIPN is believed to be the result of interactions (hydrogen and ionic bonds) between the functional groups of the two polymeric components. Finally, the differences revealed in the behaviors of the calorimetric and mechanical glass transition temperatures are regarded as experimental evidence for the existence of locally heterogeneous relaxation motions, which are probed by the different length scales associated to the calorimetric and mechanical techniques.

Thermoplastic AIPNs based on anion-containing PU_1 and polyaminourethane (PU_2) were prepared and investigated using different techniques by Stepanenko et al. [224] and Tsonos et al. [225, 226]. The PU_1 was based on oligotetramethylene glycol (OTMG, MM 1000), TDI (mixture of 2,4- and 2,6-isomers in the ratio 65 : 35), and the magnesium salt of p-hydroxybenzoic acid (Mg-HBA); the molar proportion of the reagents was 1 : 2 : 2. The PU_2 was based on the same OTMG and TDI, and N-methyldiethanolamine (NMDA); the molar proportion of reagents was 1 : 2 : 2, too. Polymer blends containing 5, 10, 30, and 50 mass % of PU_1 were obtained and studied. The details of preparation conditions are described in [224, 225]. The molecular mobility, microphase morphology, and their dependence on the composition of the blends were investigated by TSDC, DSC, FTIR, SAXS, and other techniques.

Dielectric TSDC measurements [225] of the blends have shown four relaxation mechanisms. The subglass secondary γ relaxation (at ~ 120 K) is associated with local motions of parts of the molecular chain. The β relaxation (at ~ 160 K) is attributed to the motions of the polar carbonyl groups of the polymer chain. A systematic change of the magnitude (maximum of the current) and position (temperature of the maximum current) of these two relaxations dependent on the composition of the ionomer (PU_1 content) has not been observed.

The α relaxation is related to the glass transition, T_g, of the amorphous soft phase of the PU_1 and PU_2. The Maxwell–Wagner–Sillars (MWS) relaxation is related to interfacial polarization caused by the motion of ions released during the glass transition. The temperature at which α and MWS relaxations (T_g and T_{MWS}, respectively) appear in the TSDC plots and the magnitude of these relaxations are affected strongly by the degree of microphase segregation of the PU_1 and PU_2 in the blends. Based on TSDC measurements, a parameter m_{TSDC}, a criterion expressing the relative degree of phase mixing, was introduced [225]. This parameter takes into account all the factors affecting phase mixing that the partial parameters express. According to the definition of m_{TSDC}, higher values of m_{TSDC} correspond to higher degrees of phase mixing of the components in the blends. As is seen in most of the work on AIPNs, authors mainly describe their properties. No general conclusions were drawn on the specific features of these systems.

4.5
Features of Temperature Transitions

For some systems a comparison of the properties of full and semi-IPNs has been made [227–229]. The main difference in dynamic mechanical properties consists in that the shift of temperature transitions in semi-IPNs is higher as compared with full IPNs. This effect was explained by the more complete phase separation in semi-IPNs as compared with full ones.

The transition behavior of PU–PMMA IPNs was studied by means of DSC, DMS, and thermally stimulated depolarization [230]. Instead of a sharp discontinuity in heat capacity, observed in individual networks, a very broad change was observed by calorimetry along with some intermediate changes in heat capacity. DMS yielded two peaks, which showed a marked inward shift with significant broadening. The data confirmed the two-phase structure, although the authors suggest that there is also some mutual solubility of both components due to interpenetration. When the second component is not cross-linked, the phases appear to be less entangled. Of course, the effects observed depended on the composition of the system and the amount of cross-linking agent.

The glass transitions were thoroughly investigated using the dynamic mechanical method for IPNs based on castor oil PU and epoxide–episulfide resins [102]. Again, it was found that in the IPN, the glass transitions corresponding to the IPN components shift significantly inward compared with those for pure components. The extent of inward shifting of T_g in IPN systems changes with the epoxide–episulfide ratio and composition and shows complex behavior. The authors believed that the broad glass transition peak indicates that the IPNs are partially mixed at the molecular level. To characterize the glass transition temperature shift, the authors introduced the value M defined as

$$M = \frac{\left(T_{g2} - T_{g1}\right) - \left(T'_{g2} - T'_{g1}\right)}{T_{g2} - T_{g1}}, \qquad (134)$$

where T_{g1} and T_{g2} are the glass transition temperatures of two pure components and T'_{g1} and T'_{g2} are the glass transition temperatures of the two phases in the IPN. It was discovered that the smaller is the difference between the gelation times of both networks, the higher is the value of M. The larger the extent of glass transition temperature shift, the more homogeneous is the morphology. This result indicates that the gelation time difference of the two components plays an important role in enhancing the mixing of two components in the IPN. In the analysis of their results, the authors assume that the two components of the IPN system are compatible before gelation occurs, and, when one component gels, the phase separation occurs immediately. Unfortunately, no evidence for this assumption was presented. From the glass transition temperatures, the compositions of the two phases were calculated. It was found that with decreasing difference in gelation time, the compositions of the two phases become closer and when the difference is zero, the system becomes one phase. This result seems to be rather unclear because the phase state is determined first of all by thermodynamic interactions and not by reaction kinetics.

The structural heterogeneity in a series of IPNs, poly(propylene oxide)-based PU and BMA–triethylene glycol dimethacrylate copolymer, were studied in [231] for various compositions. The data were compared with results for the pure constituent networks. To investigate the heterogeneity the method of segmental relaxation around T_g was used with the application of the laser-interferometric CRS technique. The spectra obtained over the temperature range from 150 to 360 K allow detailed characterization of the heterogeneity of segmental dynamics within or near the extraordinarily broad glass transition range in these IPNs. The systems under investigation do not show distinct microphase separation and are characterized by a single very broad glass transition. In this work for the first time a number of relaxations within or near the T_g range were differentiated. The discrete responding of the creep rate spectrum to each of these relaxations, associated with cooperative, partly cooperative, or noncooperative Kuhn segment movements, has been shown. Combining CRS data with some DSC results has allowed the molecular assignment of the multiple creep peaks to be done. The data distinctly show the nanoscale compositional heterogeneity associated with relaxations in the neat networks or segmentally mixed nanodomains. The data also confirm the incomplete compatibility for all IPN compositions.

A theoretical description of the glass transition behavior was developed for miscible IPNs and was based on the application of various equations connecting the glass transition temperature of a blend with the corresponding temperatures of pure components. A review of these approaches was given by Sperling [166]. Although this description relates only to miscible networks, it may be used for the description of the behavior of each separate phase. Sperling was the first to say that the values T_g in immiscible IPNs may be considered as the glass temperatures of a homogeneous phase in a phase-separated material. Presently, there are many equations describing the glass transition temperature for a miscible polymer blend via the glass transition temperatures of the constituent components and their volume fraction [133, 232, 234]. These equations cannot be directly applied to IPNs, as they do not take into account the effect of cross-linking on the position of T_g. This effect was considered by Frisch and coworkers [235, 236]. If T_g is the glass transition temperature for the cross-linked polymer, T_{g_0} the same for the uncross-linked polymer, and $\varepsilon_x/\varepsilon_m$ is the ratio of the lattice energies for cross-linked and uncross-linked polymer, then, taking into account the mole fraction of monomer units which are cross-linked, χ_C, and the ratio F_x/F_m of segmental mobilities for the same two polymers, the equation may be written:

$$\frac{T_g - T_{g_0}}{T_{g_0}} = \frac{\left(\varepsilon_x/\varepsilon_m - F_x/F_m\right)\chi_C}{1 - \left(1 - F_x/F_m\right)\chi_C}. \tag{135}$$

For chemically cross-linked polymers, $\varepsilon_x/\varepsilon_m \neq 1$, and the mobility of a chemically cross-linked segment $F_x \ll F_m$, so that $F_x/F_m \approx 1$, and the equation simplifies to

$$\frac{T_g - T_{g_0}}{T_{g_0}} = \frac{\left(\varepsilon_x/\varepsilon_m\right)\chi_C}{1 - \chi_C}, \tag{136}$$

which exhibits the often observed experimental increase in T_g when χ_C is increased.

In a similar way these equations may be transformed to account for physical cross-links, or entanglements, if instead of χ_C another value of χ'_C is used characterizing the increasing physical cross-linking density caused by the interpenetration in IPNs. Unfortunately, the approaches accounting for the chemical cross-linking and entanglements contribution to glass transition temperatures developed for miscible IPNs cannot be directly transferred to phase-separated IPNs, because the characteristics of the cross-linking degree in separated phases cannot be determined.

4.6
Contribution of an Interphase to Viscoelastic Properties

An important problem in describing viscoelastic properties of IPNs consists of taking into account the properties of an interphase. There is great difficulty in establishing the properties of this region using mechanical methods, and only in very rare cases does this region exhibit its own relaxation maximum. According to theoretical and experimental estimates, the thickness of the transition layer or of the interphase between two evolved phases in polymer blends ranges from several nanometers to some tenths of nanometers, although their volume fraction in the matrix may reach 30–50%. In IPNs these values are 4 nm and 25% [237]. This thickness in the case of spinodal decomposition is of the order of the spinodal decomposition wavelength. The effect of thickness of interphase regions is negligible on the macroscopic level when compared to the dimension of the body; this is not the case for small regions formed by microphase separation. The contribution of the interphase should increase with growing thickness and decreasing dimensions of phase separation regions. At the same time, the increase in the segregation degree leads to the decrease in the thickness and the fraction of an interphase. A theoretical estimate of the influence of an interphase on the viscoelastic characteristics of IPNs and similar systems may be given on the basis of a model of the Takayanagi type [184, 238, 239].

Using Eqs. 129–132 and assuming that E_A is the modulus of an interphase and E_B is the modulus of the matrix, the complex modulus of a material consisting of a matrix and an interphase was calculated. Two cases were considered: (1) the glass transition temperature of the interphase is higher as compared with the matrix, and (2) the glass transition temperature of the interphase is lower. These cases correspond to the temperature dependencies of the complex modulus $E^*(\text{matrix}) = f(T)$ and E^* (interphase) $= f(T + \Delta T)$, where $\Delta T = [T_g(\text{matrix}) - T_g(\text{interphase})]$ may be positive or negative. The calculations of this kind enabled one to establish the effect of the shift in the transition temperature of an interphase on the viscoelastic functions, the effect of the dimensions of an interphase, etc. The weakness of such an approach is that the two-phase model was used for the calculations, namely, that the properties of the two phases were not distinguishable. This shortcoming is absent in the case of a three-phase model, presented in Fig. 42 [184]. The modulus of the three-phase model E^* may be calculated as follows:

$$E^* = \left(1 - \lambda\right) E_A^* + \lambda \left[\left(1 - \varphi\right) / E_C^* + \varphi / E_B^*\right] , \tag{137}$$

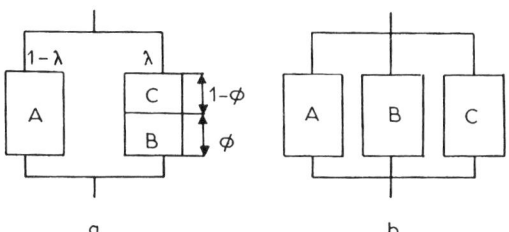

Fig. 42 Three-parameter model of two-component system with an interphase [184]

where E_A^*, E_B^*, and E_C^* are complex moduli of three phases (C is the transition layer), their volume fractions being, correspondingly: $V_A = 1 - \lambda$, $V_B = \lambda\varphi$, $V_C = \lambda(1 - \varphi)$. Parameters of the model are $\lambda = 1 - V_A + (V_C/2)$ and $\varphi = V_B - (V_C/2)$; $V_A + V_B + V_C = 1$. In Eq. 137 it was assumed that the third phase C had the character of a homogeneous mixture with the component ratio of 1 : 1 ($E_C^* = (E_A^* + E_B^*)/2$) and that V_C can assume values in the range $0 < V_C\#$ (V_{PU} or V_{PUA}). The calculated curves adequately described the experimental data (Fig. 35) for IPNs 1 and 2 using constant $V_C = 0.04$ (IPN 1) and $V_C = 0.15$ (IPN 2). However, the results for other ratios could not be adequately described using this approach.

The same assumptions about the interphase layer led to a good agreement between theoretical and experimental dependencies of E^* on T with the value V_C taken constant for the model illustrated in Fig. 42b (additivity of moduli). On the other hand, no suitable V_C values could be found for all IPNs but 1 and 2. The ability to fit IPNs 1 and 2 by both two- and three-phase models is due to a low value of V_C.

A theoretical method was proposed [240] to calculate the interphase amount in IPNs from the data on modulus–temperature behavior. For PU/PBMA IPNs the interfacial concentration was estimated and a wide interfacial region between the phase domains was discovered.

4.7
Effect of the Domain Sizes on the Glass Transition Temperatures

Some experimental data show that the dimensions of the particles of the dispersed phase affect the position of the glass transition temperature, the broadness of the transition interval, and even determine the very possibility of detecting the glass transition of this phase experimentally [241, 242].

The important question is to determine the minimum size of domains in the heterogeneous structure that allow us to detect the glass transition using conventional methods (DSC, DMS). The comparison of the electron microscopy data and methods mentioned allows us to conclude that the phase domain dimensions, at which the glass transition is seen, should be no less than 15 nm [243]. The dimensions of the phase domains correspond in this case to 200–500 carbon–carbon bonds depending on the kinetic flexibility of the macromolecules. In all cases when electron microscopy revealed microregions with dimensions less than 10 nm, only one glass transition temperature was observed. This question was studied for full IPNs based on oligoisoprene hydrazide (OIG) of different MM and EO [243]. X-ray analysis has shown that these IPNs are amorphous heterogeneous structures with an oligoisoprene matrix and rigid epoxy inclusions of very small sizes (Table 9).

The existence of two maxima on the temperature dependence curves $\tan \delta = f(T)$ (Fig. 43) for OIG MM 3850 and 2080 shows the existence of two glass transitions. For OIG with MM 1130 there occurs only one high max-

Table 9 Structural parameters and properties of IPNs based on oligoisoprene of different molecular masses and epoxy resin [236]

MM OIG	% of EO	Segregation degree	Domain dimension, nm	T_g OIG region, K	T_g EO region, K
3850	19.6	0.31	7.1	233	334
2080	31.4	0.3	6.1	253	338
1130	46.6	0.23	5.7	–	358

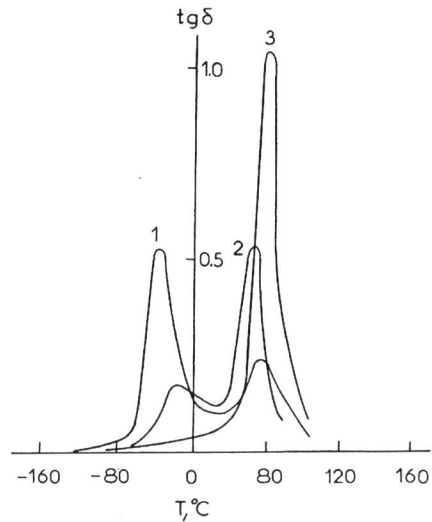

Fig. 43 Temperature dependencies of mechanical loss tangent for epoxy–diene networks. Numbers of curves correspond to (1) 19.6% of epoxy blocks, (2) 31.4%, (3) 46.6% [243]

imum, although the structural data show that the system is heterogeneous. The shift of low-temperature maxima is determined by the decreasing MM from 3850 to 2080. At the same time there is only a small shift of the high-temperature maximum for epoxy blocks. For both IPNs the glass transition may be registered. During the transition to the lowest MM of OIG, 1130, the first low-temperature maximum is not seen. The data on the segregation degree (Table 9) show that in the last case, the segregation degree is the lowest. It seems to be evident that in this case, the epoxy blocks form the continuous medium with very small inclusions of the second network, and this is the reason that the glass transition is not seen. Probably, in the latter case, the contribution of an interphase to glass transition is rather high, and some superposition of both relaxation transitions occurs, leading to the appearance of only one glass transition. The special features discussed above should be taken into account when considering the data on T_g for IPNs, indicating once more that a single transition temperature cannot be considered as evidence of compatibility of two networks in IPNs. Simultaneously, these data allow us to understand why in many cases the relaxation transitions connected with the interphase existence cannot be seen using mechanical dynamic spectroscopy.

As was said above, one glass transition point may be observed in phase-separated IPNs in the case of small domains of one of the phases. This phenomenon is very important when considering compatibilization of IPNs by introducing special substances acting as compatibilizer. For compatibilized IPNs the effect of compatibilization leads to an essential change in the viscoelastic properties. For some IPNs containing various compatibilizing agents the viscoelastic properties have been studied in [244]. It was shown that the initial IPNs are characterized by two mechanical loss regions and represent two-phase systems within a wide composition range. The action of compatibilizing additives introduced during the formation of the systems was estimated according to the change in the positions and heights of relaxation maxima or the dependence of the mechanical losses on temperature. Depending on the chemical nature of the compatibilizers, they exert a different influence on the viscoelastic functions.

4.8
Relaxation Spectra of IPNs

It is known that the most fundamental characteristic of relaxation behavior is the relaxation spectrum, which may be calculated from the following equation:

$$G(t) = G_e + \int_{-\infty}^{+\infty} H e^{t/\tau} \, d(\ln \tau), \qquad (138)$$

where $G(t)$ is the shear modulus at time t, G_e is the limiting value of the shear modulus, H is the distribution function of relaxation times, and τ is the mean relaxation time. Using the generalized frequency dependence of the real part of the complex shear modulus, the relaxation spectra may be calculated using the method proposed by Ninomia and Ferry [244]. The first relaxation spectra were calculated for sequential IPNs based on cross-linked PU and styrene–DVB copolymer [192] from the data on the temperature and the frequency dependence of viscoelastic functions (Fig. 44). It is seen that the introduction of the second network influences the relaxation spectra. In the region of small concentrations of the second network, the number of relaxators with lower rigidity and higher mobility increases, possibly due to the increasing concentration of the loosely packed interphase. Increasing the amount of the high-modulus component increases the fraction of less mobile relaxators. Thus, the effect of the first network on the molecular mobility of the chains of the second network may be explained in the same way as the effect of filler

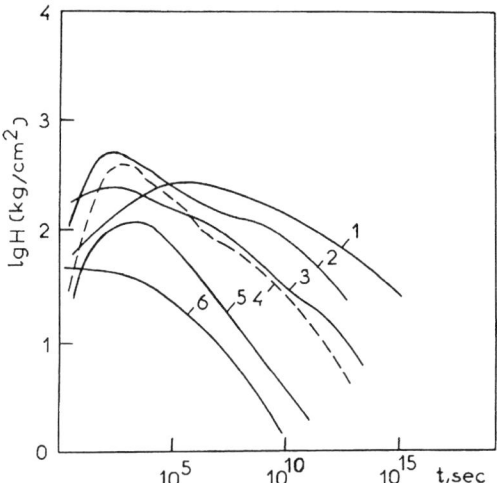

Fig. 44 Relaxation spectra of styrene–DVB/PU IPN at various volume fractions of PU: (1) 0.24; (2) 0.15; (3) 0.10; (4) pure first network; (5) 0.05; (6) 0.03 [192]

on the molecular mobility. The data on the change of molecular mobility in polymers filled with polymeric fillers are reviewed elsewhere [180].

A detailed study of relaxation behavior was done for full simultaneous IPNs based on cross-linked PU and weakly cross-linked PBMA in a wide frequency range [245]. PU was synthesized from TDI, POPG, and TMP using DBTDL as a catalyst. Monomeric BMA and cross-linking agent dimethacrylate triethylene glycol with dissolved initiator (AIBN) were introduced into the reaction mixture. Two IPNs were prepared with ratios of two networks PU/PBMA of 65 : 35 and 50:50 mass %. Figures 45 and 46 represent the temperature dependencies of the dynamic shear modulus G' and the mechanical loss tangent δ for the studied IPNs. It is seen that the increasing frequency leads to a shift of the temperature dependence of the shear modulus to higher temperatures. The glass transition temperature for PU increases with the frequency from 225 to 251 K and for PBMA from 313 to 338 K (Fig. 45a,b). More complicated plots of log $G'(T)$ and tan $\delta(T)$ are observed for IPNs (Fig. 45c,d and Fig. 46c,d). For the 50 : 50 IPN with an increase in frequency the glass transition temperature of the PU-enriched phase is shifted from 223 to 233 K, for the 65 : 35 IPN, from 233 to 248 K, and for the PBMA-enriched phases, from 303 to 325 K and from 315 to 343 K, respectively. For the 65 : 35 IPN, as distinct from the 50 : 50 IPN, there is also observed a third, intermediate maximum, or shoulder near the maximum, corresponding to the phase enriched in PU (Fig. 46c). This maximum may be attributed to the interphase region between the two evolved phases. It was impossible to trace the shift of this intermediate maximum with frequency because at higher frequencies this maximum degenerates into a shoulder (Fig. 46c, curves F and G).

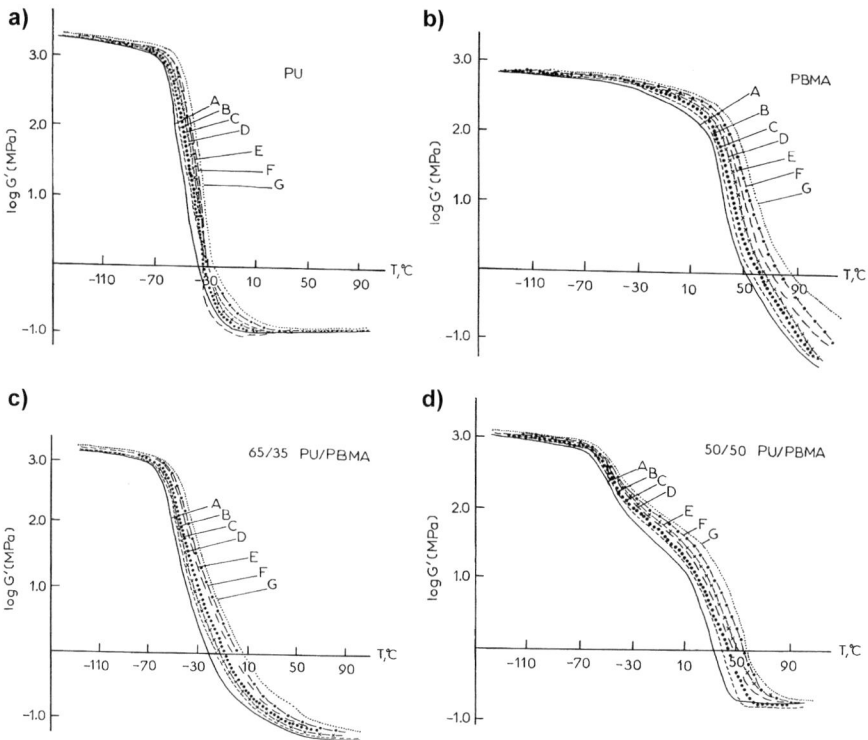

Fig. 45 Temperature dependence of dynamic shear modulus for initial networks (**a,b**) and for IPNs with component ratio 65 : 35 (**c**), 50 : 50 (**d**) by mass % at various frequencies (Rad s⁻¹): A 0.1, B 0.31, C 1.0, D 3.1, E 10.0, F 31.1, G 100.00 [245]

For the first time in the cited work the frequency dependence of the segregation degree, calculated from the parameters of relaxation maxima, was studied (Table 10) [245]. As is seen from Table 10, the changing frequency does not influence the values of segregation degree estimated from relaxation maxima. The comparatively low values of the segregation degree show that a great amount of the system is preserved in the unseparated state; this means that the evolved phases are not pure components but mixtures of both networks.

From the temperature dependence of tan δ it follows that for the 65 : 35 IPN the continuous phase is the PU-enriched phase, whereas for the 50 : 50 IPN the continuous phase is PBMA-enriched phase. From the temperature dependence of tan δ at various frequencies the activation energies of the relaxation processes in IPNs have been calculated [192] using the Arrhenius equation:

$$\tau = A \exp\left(- E_{a}/RT\right) . \tag{139}$$

Here τ was found from the condition $\omega\tau = 1$ for the maximum tan δ (ω is the frequency). Because IPNs have two glass transition temperatures, it is pos-

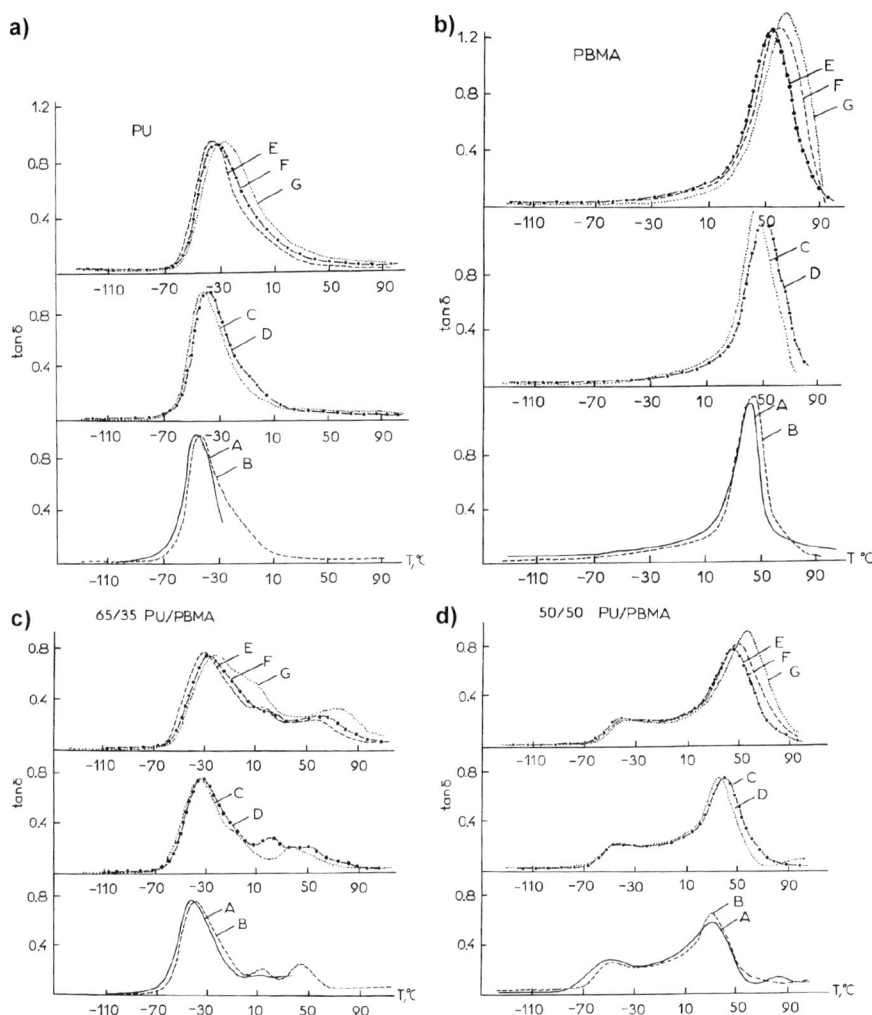

Fig. 46 Temperature dependencies of mechanical losses for initial networks (**a,b**) and for IPNs with component ratio 65 : 35 (**c**), 50 : 50 (**d**) by mass % at various frequencies (as in Fig. 45) [245]

sible to calculate the activation energy for each phase separately. The data are presented in Table 11.

It can be seen that for each IPN the values of E_a are practically the same for both phases. For the 65 : 35 IPN, E_a for both phases is very close to E_a for pure PU, whereas for the 50 : 50 IPN, E_a is different for the two phases and only for the PBMA-enriched phase is E_a the same as for pure PBMA. These results seem to be rather puzzling and may be connected to the dual-phase morph-

Table 10 Frequency dependence of the segregation degree [245]

Frequency, Rad s^{-1}	Segregation degree, α	
	65 : 35 IPN	50 : 50 IPN
0.10	–	0.36
0.31	0.38	0.37
1.0	0.39	0.40
3.1	0.31	0.38
10.0	0.34	0.44
100.0	0.36	0.43

Table 11 Activation energies of relaxation in the pure networks and in IPNs [245]

Network	Activation energy E_a, kJ·mol^{-1}	
PU	180 ± 15	
PBMA	280 ± 25	
IPN, network ratio	PU-enriched phase	PBMA-enriched phase
65 : 35	170 ± 15	180 ± 15
50 : 50	350 ± 30	290 ± 25

ology and phase inversion; however, the authors of the cited work presented no morphological data.

The analysis of the relaxation spectra below shows that IPNs with longer relaxation times (50 : 50) have higher activation energy in each evolved phase. The continuous changes in the relaxation spectrum do not allow for the distinction between those parts of the spectrum belonging to different phases. However, the very existence of phase continuity cannot explain the equality of the activation energies. The data on activation energies probably show that the structures of the two evolved phases are very different for IPNs of various compositions. This difference in the relaxation behavior seems to be connected to different correlations between quasi-independent and cooperative segmental motions in each phase, in accordance with the concept developed by Berstein [246]. Again, it is worth noting that the two phases are not in a state of true thermodynamic equilibrium and may be considered as dissipative structures [247].

The characteristics of dissipative structures in IPNs are strongly dependent on the condition of nonequilibrium phase separation. According to Berstein, in the systems with dissipative structures there is the probability of the parallel occurrence of α relaxation and β relaxation processes, the transition between them being accompanied by increases in both the cooperative degree and the activation energy of transition.

The temperature and the frequency dependencies of the dynamic shear modulus G' in the frequency range $0.1 - 100.0$ Rad s^{-1} allowed calculation of the generalized curves of the viscoelastic functions using the method of reduced variables. This method enables one to extend the frequency range [244]. The choice of the reduction temperature $T_0 = 273$ K is worth explaining. As a rule, the reduction temperature is chosen in such a way that $T_0 = T_g + 50$. As discussed above, IPNs are two-phase systems and have two glass temperatures. The choice of $T_0 = 273$ K was made because this temperature lies between the glass transitions of the two constituent networks. The generalized curves are presented in Fig. 47. The method of reduced variables has allowed the function of dynamic shear modulus to be spread over 16 orders along the frequency axis. The generalized curve may be considered as consisting of three regions. The first is the region of glassy state I, where the chain mobility is frozen. The second region (II) is the transition zone from the glassy state to the rubberlike state. The third region (III) is a plateau of rubberlike elasticity where the dynamic properties are connected with the existence of a network of molecular entanglements and chemical cross-links. Analysis of the generalized curves shows that the PU network has the greatest value of G' in a glassy state (three or four orders). For cross-linked PBMA the glassy region is spread over seven or eight orders along the frequency axis. IPNs are characterized by the intermediate values of G' in the glassy state spread over five to six orders. The most marked distinctions in the generalized curves for the IPNs are observed in the transition zone (zone II). Analysis of its position on the frequency axis shows that both PU and PBMA have rather narrow transition regions and the curves for these polymers are almost parallel. The transition zone for PU is shifted four orders to higher fre-

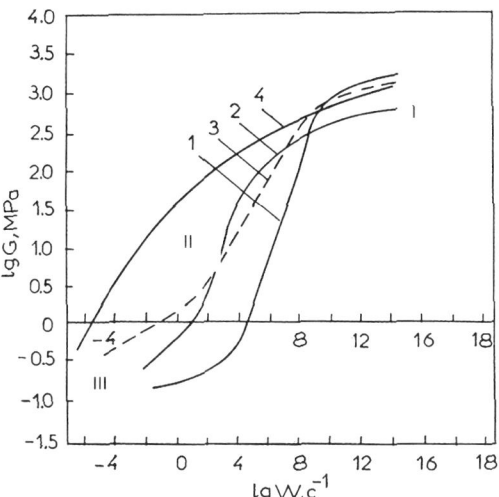

Fig. 47 Generalized dependencies of dynamic modulus reduced to 273 K: PU (1), PBMA (2), 65 : 35 IPN (3), 50 : 50 IPN (4) [245]

quencies compared with PBMA. Therefore, the relaxation processes in the PU region of the glass transition have shorter relaxation times than those processes in PBMA. The distinctive feature of IPNs is that they have very wide transition regions and therefore have a broad set of relaxation times. Qualitatively this agrees very well with the data on the free volume distribution. For the 50 : 50 IPN the transition zone is much broader than that of the 65 : 35 IPN and is spread to a lower frequency over six to eight orders. It can be concluded that the 50 : 50 IPN is characterized by relaxation processes with longer relaxation times. In such a way, changing the ratio of the IPN components led to the change in the relaxation behavior. This is very important for the damping behavior of such systems.

From the data considered above, the relaxation spectra have been calculated and presented in Fig. 48. Both the initial networks and the IPNs give a broad maximum on the relaxation spectrum, which corresponds to the most common relaxation time in each case. The region of the sharp decrease on the $\log H(\log \tau)$ curve is a typical transition region from the glassy to the rubberlike state. The end zone for cross-linked PBMA is characterized by a drop in the relaxation spectrum, whereas in the pseudo-equilibrium region a plateau is observed (curve B). For cross-linked PU (curve A) the relaxation times are rather short and the maximum on the $\log H(\log \tau)$ curve occurs over the range $10^{-12} < \tau < 10^{-8}$ s. This means that in the PU network, rapid conformational rearrangements are taking place. In the pseudo-equilibrium zone there is some sign of a plateau connected with the existence of a chemical network. For weakly cross-linked PBMA (curve B) the region of the glassy state is located at the region of shorter relaxation times. The maximum of the main relaxation transition is observed in the range $10^{-7} < \tau < 10^{-5}$ s. Comparison of the PU and PBMA spectra shows the slowness of the relaxation processes in PBMA compared with PU. At the same time, from the dependence of $\log H$ on $\log \tau$, it can be seen that the slopes of the plots for PU and

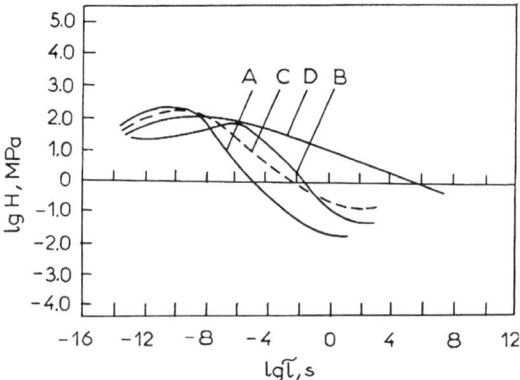

Fig. 48 Relaxation spectra: PU (*A*), PBMA (*B*), 65 : 35 IPN (*C*), 50 : 50 IPN (*D*) [245]

PBMA are practically the same. According to Ferry theory [244] this slope for homopolymers is equal to $-1/2$, which agrees with the data obtained for PU and PBMA. For the IPNs there is a sharp deviation of the slope from $-1/2$ due to microheterogeneity of the system and the broader set of relaxation times than in homopolymers. For the $65:35$ IPN the maximum on the $\log H(\log \tau)$ curve is observed in the region $10^{-12} < \tau < 10^{-7}$ s and transition from the glassy state is characterized by relaxation times in the range $10^{-12} < \tau < 1.0$ s (curve C). For the $50:50$ IPN the broad maximum occurs over the same interval and the transition region is located in the range $10^{-12} < \tau < 10^{-5}$ s (curve D). Thus, the $50:50$ IPN has a much broader set of relaxation times than the $65:35$ IPN.

In such a way, IPNs have a much broader set of relaxation times than the pure constituent networks. The relaxation spectra of IPNs cannot be obtained by simple superposition of the spectra of the constituent networks. The broader spectra of the IPNs may be explained by the existence of a two-phase structure where each phase is enriched in one of the components. Simultaneously, the existence of two phases is reflected in the relaxation spectra by their broadening and shift along the time axis. It is evident that in spite of the incompatibility of the two networks, there exists a strong physical interaction between macromolecular chains of dissimilar chemical nature, which may be described in terms of entanglements, and strong polar interactions. The latter may be the reason why $50:50$ IPN has its relaxation spectrum shifted to higher relaxation times than that for $65:35$ IPN.

By considering these data, we take into account the influence of morphological features on the dynamic mechanical properties of the two-phase system. This effect may also determine the relaxation characteristics and the activation energies. Since any relaxation process is connected with molecular mobility and interchain physical interactions, the morphology definitely affects the relaxation properties.

4.9
Vibration-Damping Properties

It is known that phase-separated IPNs made from low and high T_g polymers damp noise and vibrations over the intervening transition range. As was pointed out by Sperling [166], the movement of flexible chains over their stiffer neighbors may underlie this phenomenon, which is closely connected with the viscoelastic behavior and relaxation spectra of the material. According to Sperling, when polymers are in their glass transition region, the time required to complete an average coordinated movement of the chain segments is approximately the same as the measurement time. If dynamic or cyclic motions are involved, the time required to complete one cycle (or its inverse, the frequency) becomes the important time unit. In the glass transition region the conversion of mechanical energy into heat reaches its maximum value. The conversion of vibrational energy to heat occurs when a polymer at its glass transition temperature is in contact with a vibrating surface. Extensional damping is caused by the loss modulus G''. For certain simple types of deformations, the heat generated per unit volume per cycle is

$$H = \pi E' \varepsilon_0^2 , \tag{140}$$

where ε_0 is the maximum amplitude. Here we also assume that Hooke's law is valid. When vibration-damping polymers are created, it is desirable to maximize E''; however, there is a limit to it, because maximizing E'' may lead to a lower elastic response and, therefore, to poor mechanical properties.

The main factors that affect the damping capability of polymers were reported elsewhere [248]. The theory of frequency-dependent attenuation of vibrations by viscoelastic polymers was presented [249]. In review [55] there are some data on dynamic mechanical properties and on the damping characteristics, which are tightly connected.

To quantify damping performance, the area under the $\tan \delta$ or E'' vs temperature curves may be used. The loss area is given [250] as:

$$\mathrm{LA} = \int_{T_g}^{T_R} E'' \, dT = \left(E'_g - E'_R \right) R / \left(E_a \right)_{\mathrm{avg}} [\pi/2] \, T_g^2 , \tag{141}$$

and

$$tA = \int_{T_g}^{T_R} \tan \delta \, dT = \left(\ln E'_g - \ln E'_R \right) R / \left(E_a \right)_{\mathrm{avg}} [\pi/2] \, T_g^2 , \tag{142}$$

where E'_g and E'_R are the storage moduli in the glassy and rubbery states, respectively, T_g and T_R are the glassy and rubbery temperatures just below and above glass transition, $(E_a)_{\mathrm{avg}}$ is the average activation energy of the relaxation process, and R is the gas constant.

To predict the damping behavior, a group contribution analysis was developed [251]. It is based on the assumption that the integral of loss modulus versus temperature curve characterizes a relation between the extent of damping and the contribution from various chemical groups of a polymeric system. The group contribution analysis for the loss area, LA, is based on the assumption that the structural groups in the repeating units provide a weight fraction additive contribution to the total loss area. The basic equation for the group contribution analysis of LA is [250]:

$$\mathrm{LA} = \sum_{i=1}^{n} \frac{\left(\mathrm{LA} \right)_i M_i}{M} - \sum_{i=1}^{n} \frac{G_i}{M} , \tag{143}$$

where M_i is the molecular mass of the i-th group in the repeating unit, M is the MM of the polymer, G_i is the molar loss constant for the i-th group,

$(LA)_i$ is the loss area contributed by the i-th group, and n represents the number of moieties in the polymer. When one attempts to calculate the $(LA)_i$ contributions, a series of simultaneous equations results with one more unknown than is the number of equations. The extra unknown is the backbone structure common to all polymers. The family of equations has the general form:

$$LA_p = w_b(LA)_b + \sum w_i(LA)_i, \tag{144}$$

where LA_p is the LA value for the polymer in question, the subscript b represents the backbone, and the subscript i represents all other moieties in the polymer. The quantity LA_b may be determined either by a trial and error method until consistent values are obtained throughout the family of polymers, or through the study of polymers with long aliphatic side groups. Both these methods were used in [250]. A method of evaluating of the area under the linear loss modulus–temperature curves was developed [252, 253], which accounts for the background. The method proposed was applied to studying various sequential IPNs (poly(methyl acrylate) (PMA)/PMMA, PMA/PHEMA, PBMA/PS, PVME/PS, and others). It was suggested that for IPNs an additive mixing rule could be written:

$$LA = w_I(LA)_I + w_{II}(LA)_{II}, \tag{145}$$

where w_I and w_{II} are the mass fractions of components I and II in IPNs. The weakness of this approach consists of its inapplicability to phase-separated systems, because they have two loss tangents and because usually the compositions of the two phases are not known. It is evident that each loss maximum in the phase-separated IPN is determined by the group contributions of both networks and by their ratio in the phase. If these factors were taken into account, the method could be very useful for estimating the damping behavior of phase-separated IPNs. It is the existence of two loss peaks that enables estimation from the parameters of these peaks (height and width) the segregation degree in phase-separated IPNs.

4.10
Determining the Segregation Degree
from Parameters of Relaxation Maxima

It was already mentioned that the properties of the polymer blends and alloys are determined by their microphase structure. The microphase state may be characterized by the degree of the microphase separation (segregation degree) and by the size and distribution of microregions of phase separation. These characteristics are connected to the history of the system. The system with uncompleted phase separation is characterized by the segregated structure with nonequilibrium regions of microphase separation. These regions may have different compositions, densities, and sizes.

Sperling [166] proposed a method for the estimation of the degree of incompatibility by introducing the concept of the "incompatibility number" N, which may be defined as

$$N = 1 - \frac{X_1^{-1} + X_3^{-1}}{2X_2^{-1}}, \tag{146}$$

where $X = d \log E / dt$, indices 1 and 3 correspond to two transition regions, and index 2 to the plateau between transitions. The value N changes in the interval between zero (compatible system) and 1 (incompatible system). This method was used, in particular, to characterize the PU/PDMS IPNs with different compositions (Fig. 49) [254]. As is seen, for a given system the incompatibility numbers are at a minimum for the IPNs at the inversion point, which suggests that the maximum compatibility of the networks occurs around the inversion point. Another way to estimate the segregation degree is based on analysis of the relaxation maxima of the system. In this case, again, the structure of the system may be characterized by a single fundamental parameter [255].

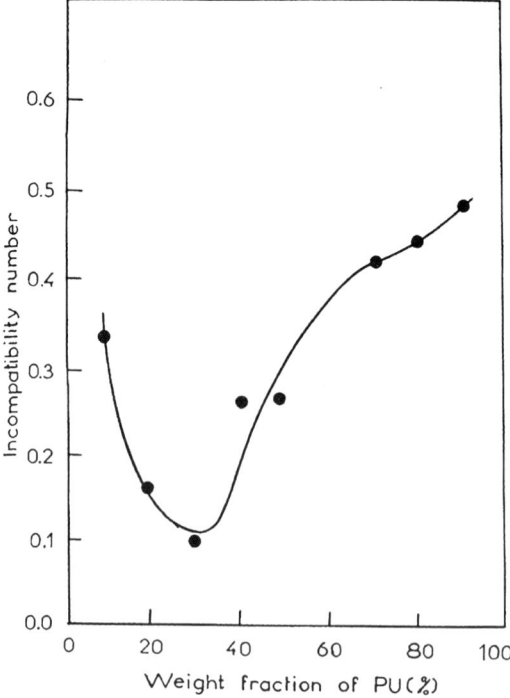

Fig. 49 Incompatibility numbers of the PU/PDMS IPNs [254]

A scheme of the microphase separation process and the determination of the degree of segregation may be presented as follows. The scheme is based on the estimation of the tangent of mechanical losses. As is known, its maximum is in the area of relaxation transition and changes at the glass temperature by more than an order of magnitude. It is also known that the polymer–polymer immiscibility is indicated by the appearance of two or more maxima of mechanical (or dielectric) losses. Let us consider schematically the temperature dependence of mechanical losses in the two-phase polymer system with different degrees of component segregation (Fig. 50). The diagram is idealized and can be described as follows.

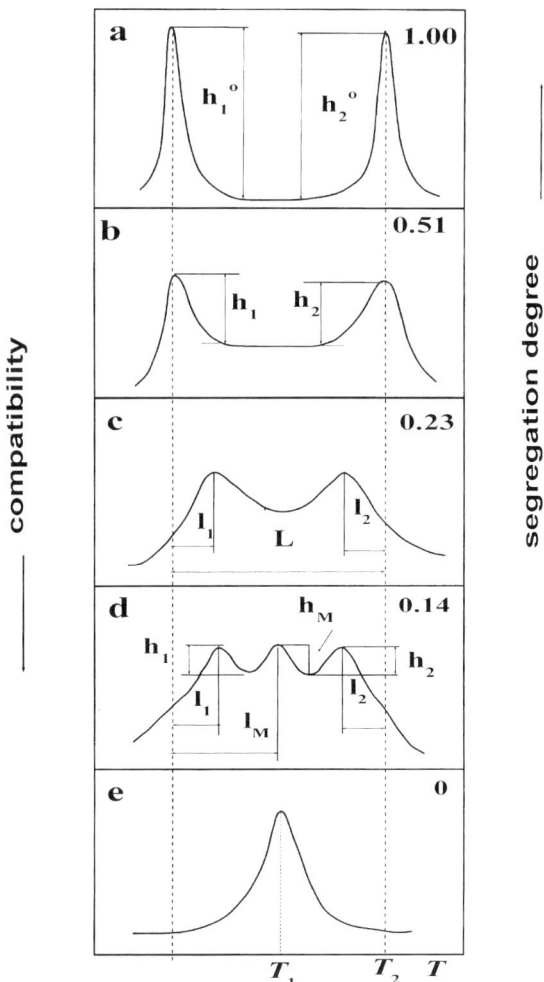

Fig. 50 a–e Schematic representation of the temperature dependencies of mechanical losses with different degrees of component segregation and compatibility [255]

Case a corresponds to a mixture of two components where the phases are clearly separated and each one is characterized by its own glass temperature. An increase in the component interaction and their mixing causes a decrease in the maxima by their absolute values and the convergence of the maxima (b and c). The compatibility level obtained after a stable interphase region has been formed is characterized by a glass transition temperature T_g between T_1 and T_2. Eventually, an increase in the interaction gives rise to a strong maximum, which tends to converge when compatibility on the molecular level is attained, i.e., when the one-phase system is formed. A reverse sequence of changes will be observed by the transition from the initial one-phase system to the two-phase system with incomplete separation. An analytical characteristic of the degree of segregation may be obtained as follows.

$\alpha = 1$ corresponds to the case of complete separation, $\alpha = 0$ to that of complete mixing. Numerous experimental data indicate that the absolute value of max tan δ is more sensitive to structural changes than the shift across the temperature scale. The following equation may be written, then, for case b considering only a decrease in the absolute value:

$$\alpha = \left(h_1 + h_2\right) / \left(h_1^0 + h_2^0\right) , \tag{147}$$

where h_1^0 and h_2^0 are the values of mechanical losses for pure components (case of complete phase separation), while h_1 and h_2 are the values for each component at different degrees of segregation.

These changes may be considered as changes in the number of relaxator units (segments) participating in the cooperative process of glass formation at the given temperature. A decrease in h for phase 1 indicates that the number of these units has decreased due to their interaction with the relaxator units of phase 2, and vice versa. The maxima may expand in this case.

Next we consider the convergence of max tan δ when the compatibility increases. To this end, we introduce the empirical parameter λ_T into the expression for α, considering the shift of the maximum across the temperature scale for phases 1 and 2:

$$\lambda_T = \frac{l_1 h_1}{L} + \frac{l_2 h_2}{L} , \tag{148}$$

where l_1 and l_2 are the shifts of the maxima across the temperature scale and L is the interval between the glass temperatures of the pure components.

The appearance of a relaxation maximum in the interphase region is accounted for by using another parameter:

$$\lambda_m = \frac{l_m h_m}{L} , \tag{149}$$

where l_m is the maximum shift of the interphase region across the temperature scale as related to phase 1. The following expression may be written in this case for α:

$$\alpha = \frac{h_1 + h_2}{h_1^0 + h_2^0} \left(\lambda_T + \lambda_m \right) . \tag{150}$$

The negative contribution of the parameters λ_T and λ_m is evident from physical considerations, since the shift of $\max \tan \delta$ and the development of an interphase layer is a result of a decrease in the segregation degree. The final expression for calculating the degree of segregation is:

$$\alpha = \frac{\left[h_1 + h_2 - \left(l_1 h_1 + l_2 h_2 + l_m h_m \right) /L \right]}{\left(h_1^0 + h_2^0 \right)} . \tag{151}$$

Calculations using these correlations produce the following data for the diagram in Fig. 50: (b) $\alpha = 0.5$; (c) $\alpha = 0.23$; (d) $\alpha = 0.14$.

The comparatively low values of segregation degree show that a large portion of the system is preserved in the unseparated state and is distributed in two phases as an interphase. Thus, segregation degree is a measure of incompleteness of the phase separation or of deviation of the system from the state of true thermodynamic equilibrium. It should, however, be noted that the model considered relates only to the two-phase system and does not account for the possible appearance of an interphase between two evolved phases.

4.11
Dependence of Viscoelastic Properties on Segregation Degree

From what was said about the reaction kinetics of IPN formation it follows that the segregation degree depends on the ratio of rates of chemical reactions and, therefore, on the conditions of phase separation. We may think that the viscoelastic properties will be dependent on the reaction kinetics and segregation degree, which, in its turn, is also determined by reaction conditions. This represents a great difference in describing the properties of IPNs as compared with the properties of blends of linear polymers.

The general picture of viscoelastic properties was considered when we discussed the method of determining the segregation degree. The majority of cases may generally be described by this scheme. However, it was shown that the viscoelasticity depends not only on composition, but also on the order of the network formation and the method of synthesis. All this is easily explained if we remember that all these factors affect the segregation degree.

The dependence of the viscoelastic properties of IPNs based on PU and PUA on the segregation degree was shown in [88, 256, 257]. Taking into account the interconnection between the chemical kinetics and the kinetics of phase separation, various methods of IPN synthesis have been used, i.e., simultaneous and sequential methods. It is known that oligourethane acrylate (OUA) in the presence of photoinitiators polymerizes at a high rate at room temperature, the reaction rate being much higher than that of the polyaddition reaction. The PU component of the IPN consisted of POPG MM 600 and adduct based on 2,4-TDI and TMP. OUA was synthesized from POPG MM 2000, 2,4-TDI, and HEMA. Variations in the degree of segregation were achieved in several ways:

1. Change in the sequence of network curing: (a) first stage—PU curing (7 days), second stage—OUA polymerization (2 h); (b) first stage—OUA polymerization (2 h), second stage—PU curing (7 days); both stages were performed at 293 K.
2. Simultaneous curing of both networks at 373 K for 2 h.
3. Polymerization of OUA at various stages of PU curing (photoinitiation after 4, 20, 64, and 100 h).

The experimental data on the temperature dependence of the loss modulus E'' show that all the IPNs have two glass transition temperatures. However, the sequence of network formation contributes to the microphase separation. The segregation degree is higher when the PU network is cured first ($\alpha = 0.29$), whereas when OUA is polymerized first, $\alpha = 0.11$. When PU is formed first, the change in curing time does not influence the glass temperature of the PU phase, the segregation degree being small as well (0.29–0.32). However, when OUA is polymerized first, the increase in the rate of PU curing (at 373 K) sharply increases the degree of segregation (from 0.11 to 0.21). When polymerization begins during PU curing (various conversion degrees of isocyanate groups), the dependence of segregation on the isocyanate conversion goes through a maximum (Fig. 51). Such behavior may be explained from the point of view of the structural theory of gelation [23].

In the first stage of PU formation, some microgel particles are formed which serve as nuclei for the three-dimensional network formation and lead to the microseparation of the PU system. Thus, the OUA network is formed

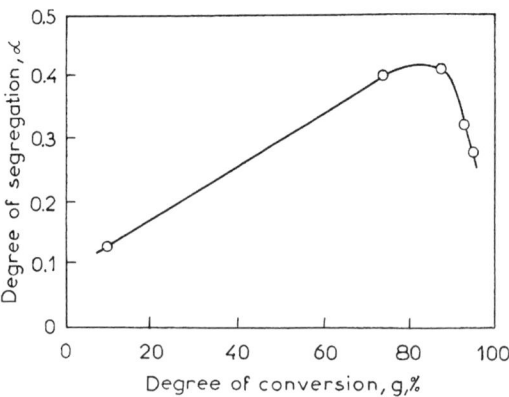

Fig. 51 Dependence of the segregation degree in IPNs on the conversion degree in the PU network [88]

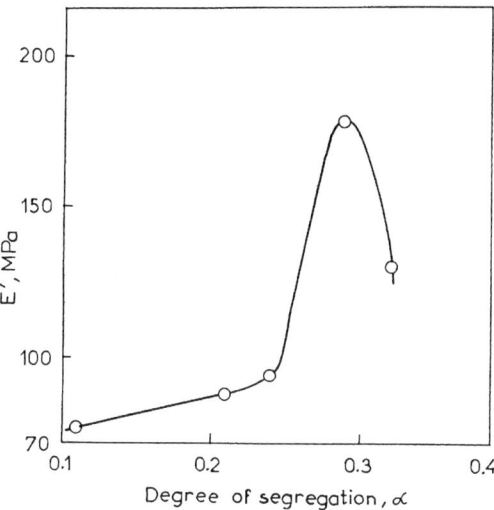

Fig. 52 Dependence of elastic modulus, E', on the degree of segregation [88]

in the microheterogeneous PU matrix. In this case, the degree of microphase separation will depend on the size of the microgel particles and on their connection by the tie chains. With the growth in microgel particle size, the degree of segregation increases up to the point where a PU macrogel is formed. The greatest degree of segregation is reached when PUA network formation begins at the stage of PU macrogel formation. When polymerization of OUA begins at higher isocyanate conversion degrees, the degree of segregation diminishes due to the hindrances imposed by the PU matrix on the phase separation.

Figure 52 shows the dependence of the elastic modulus E' on the degree of segregation, demonstrating its interconnection with mechanical properties. The decrease in the modulus at higher degrees of segregation (above 0.3) may be explained by an increase in the number of network defects [88]. Another attempt was made to relate the conditions of microphase separation (determined by curing temperature) with the viscoelastic properties using a semi-IPN based on styrene–DVB copolymer and PBMA. Figure 53 shows the typical temperature dependence of tan δ for various semi-IPNs containing 40 mass % of PBMA and synthesized at various temperatures. Each maximum is located near the temperature range of the tan δ maximum for pure components. Some shift of T_g is the result of each phase consisting of both components.

It was shown that the increasing curing temperature increases the degree of segregation, i.e., the completeness of phase separation in the system. For the system under consideration, the phase separation proceeds according to the spinodal mechanism. With an increase in the curing temperature, the rate

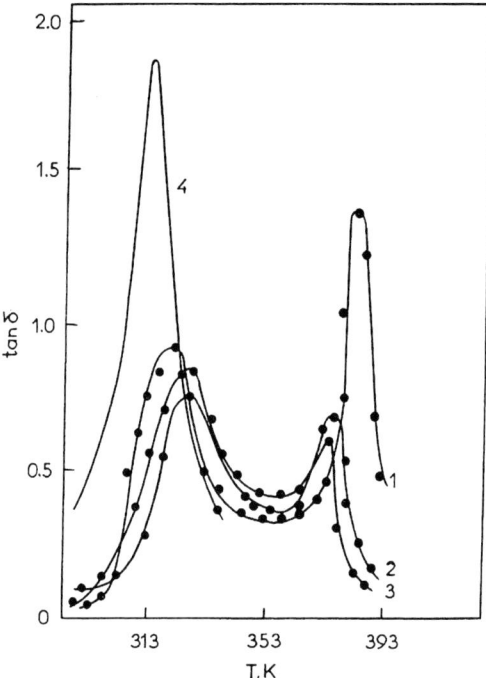

Fig. 53 Temperature dependence of $\tan\delta$ for sequential IPNs made from styrene–DVB copolymer and PU and synthesized at various temperatures: (1) 363 K, (2) 353 K, (3) 333 K, (4) pure PBMA [88]

of phase separation markedly increases. Correspondingly, the values of segregation degree are higher for samples synthesized at elevated temperatures. However, the segregation degree in all cases is not very high, being considerably lower than 1. For semi-IPNs made from cross-linked PU and PBMA it was also established that, depending on the reaction rates of formation of the two polymers, the segregation degree may change in a remarkable range—from 0.11 to 0.54.

Thus, a general conclusion can be drawn that the viscoelastic properties of the full and semi-IPNs depend not only on the miscibility and thermodynamic interaction between two constituent networks, but also on the reaction kinetics. Both thermodynamic and kinetic factors determine the viscoelasticity of these systems due to superposition of the chemical and physical processes occurring in the systems during curing.

5
Chemical Kinetics of IPN Formation and Phase Separation

5.1
General Premises

As distinct from almost all polymer materials, the kinetics of IPN formation is a governing factor in development of the system morphology. For traditional polymeric materials their structure and morphology depend on the ways of processing, heat treatment, and other physical, not chemical, factors, while for IPNs all depends on the kinetic conditions. One may say that thermodynamics gives the general rules of the equilibrium state and determines the path to equilibrium, whereas the kinetics allows the realization of the path and predetermines the real structure far from equilibrium. This specific feature of the IPN formation is connected with the fact that in the reaction system two processes proceed simultaneously: the chemical process of network formation and the physical process of phase separation. As will be shown below, these processes are interconnected.

By considering the reaction kinetics of the individual network or linear polymer formation in IPNs, one should also bear in mind that the data on the kinetics obtained for pure components could not be used to describe the IPN synthesis for three main reasons [51]:

1. In the case of sequential IPNs the reaction from the very beginning proceeds by definition in the medium of the first matrix networks. In the case of simultaneous IPNs the network that is formed earlier, due to the difference in reaction kinetics, also forms the medium for the development of the second network and serves as a matrix. The matrix network changes the reaction conditions at the expense of the variation in the ratio of the rates of elementary reactions (chain propagation, transfer, and termination), and due to the possibility of chain transfer or termination on the chains of another network leading to grafting.

2. The matrix network changes the diffusion parameters of the reaction and influences the reaction proceeding in the diffusion region.

3. In the course of the synthesis of both the sequential and the simultaneous networks, microphase separation of the system occurs as a result of the appearance of immiscibility of growing chains of both network components. During IPN synthesis the kinetics of the formation of each network is different, primarily because of different formation mechanisms—in general, these mechanisms are polymerization and polyaddition. Usually one network is formed first and serves as a matrix for the formation of the second network.

The following effects are operating here—a total change in the viscosity of the reaction medium, which determines kinetics, and possible physical interaction between various groups of growing chains of both components. Both effects, thus, are influencing the reactions proceeding in the diffusion region [51]. These three principal factors lead to the differences in the kinetics of IPN formation as compared with individual networks. From what was said above it is clear that the kinetics of IPN formation could not be considered without simultaneous analysis of the phase separation during synthesis. Indeed, during both simultaneous and sequential formation of IPNs, the initial reaction system (mixture of components for both networks and swollen gel) is a one-phase system. In the course of reaction, the thermodynamic incompatibility of growing networks appears at a certain value of conversion degree. Because of this, the microphase separation is determined by the rate with which the system achieves the state of immiscibility. Thus, the properties and morphology of IPNs are determined by the reaction kinetics. One can say that thermodynamics governs the general rules of reaching the equilibrium state, whereas the reaction kinetics determines whether the equilibrium is possible, and if it is, to what degree.

Despite the great importance of studying the kinetics of IPN formation, only two reviews on this subject were published [258, 259]. In the present chapter the general characteristics of IPN formation will be considered within the framework described above.

5.2
Kinetic Characteristics of IPN Formation

One of the first investigations of the kinetics of IPN formation was done [165] for the IPNs made of epoxy and PnBA cross-linked by diethylene glycol dimethacrylate. The kinetics of formation of the components was described within the framework of traditional approaches.

The kinetics of IPN formation was studied for the system PU/cross-linked PMMA [260]. PU was formed from POPG MM 2000 and from aromatic triisocyanate adducts in the presence of catalyst (Sn octoate). The PMMA network was synthesized by radical copolymerization of MMA and trimethylolpropane trimethacrylate using AIBN as initiator. First, the PU network was formed at room temperature and then at 223 K the second network was polymerized. The effect of all components of one of the networks was studied on the gelation of the other. The effect of the catalyst Sn octoate on the PMMA gel formation consisted in a decrease in the rate of gelation from 180 to 30 min. At the same time, the PMMA network components do not influence the PU network formation. The first effect was explained by the increase of the viscosity of the reaction medium, leading to an increased rate of PMMA formation in the presence PU components due to a smaller termination constant according to the equation [261]:

$$V_g = \frac{K_g[M]\sqrt{V_i}}{\sqrt{K_t}}, \tag{152}$$

where $[M]$ is the monomer concentration, K_g is the chain growth constant, V_g is the rate of chain growth, V_i is the initiation rate, and K_t is the termination constant.

It was established [262] that the catalyst Sn octoate does not exert any effect on the initiator decomposition rate, but enhances the polymerization rate of monomer and comonomer due to polarization of double bonds, making addition of radicals easier; as a result, the rate of PMMA formation increases. The possibility of chemical bond formation between PU and the PMMA network was also discussed. One should never forget such a possibility because of the possible chain transfer to another polymer. Two components of the PU network—POPG and triisocyanate adduct—both containing mobile hydrogen atoms, are present in the reaction medium. Because this atom may be "shipped off" by a radical, the possible reaction may be presented as consisting of two stages:

I. $\quad$ Ac$^{\cdot}$ + T $\rightarrow$ Ac + T$^{\cdot}$

II. (a) $\quad$ T$^{\cdot}$ + Ac $\rightarrow$ T $\sim$ Ac$^{\cdot}$

$\quad$ (b) $\quad$ T$^{\cdot}$ + Ac$^{\cdot}$ $\rightarrow$ T $\sim$ Ac ,

where Ac is MMA monomer, T is a component containing the oxypropylene or urethane groups, and Ac$^{\cdot}$ and T$^{\cdot}$ are the corresponding radicals. If the reactions II (a) and (b) take place, the chemical bond between PU and PMMA may be formed. To prove this assumption, the mixture of MMA was polymerized with various amounts of α,ω-diphenylurethane adduct of POPG.

The reaction products were extracted by a solvent and analyzed. It was found that more than 90% of the PU component could be extracted. For the system containing 20% of PU and 80% of PMMA, after extraction there is about 1.8% of PU left in the PMMA network. This low value shows that even if the grafting really takes place, its effect on the macroscopic properties of the IPN is negligible. In the cited work it was emphasized for the first time that the analysis of the reaction kinetics should account for both the effect of one network on the formation of the second, and vice versa, and the possibility of secondary reactions that may prevent or distort the formation of one of the networks.

To study the kinetics of semi- and full IPN formation from PU and PMMA, FTIR spectroscopy was used which allowed simultaneously investigation of the consumption of NCO groups (absorption band at $2275\,\mathrm{cm}^{-1}$) and C $=$ C bonds ($1639\,\mathrm{cm}^{-1}$). The effects of various factors were established, including the temperature, cross-linking agent concentration, and PU concentration in solution, etc. It was found that the shape of the kinetic curves of PU and PMMA formation in IPNs does not change as compared to pure component, this being the basis for the conclusion about the absence of any chemical interaction between two networks. IPNs made of the same components were also investigated using stepwise polymerization [263]. PU was formed at room temperature and PMMA at 333 K.

The effect of concentration of the catalyst and of PU/PMMA ratio on the reaction rate constant for PU was investigated. By studying the PU kinetics at the component ratio value of 34:66 mass % for simultaneously proceeding reactions it was discovered that the rate of PU formation increases eight times with increasing temperature [264]. The reaction order was estimated in a broad interval of NCO conversion.

$$d\,[NCO]\,/\,dt = K\,[OcSn]\,[NCO]^{1.5}\,. \tag{153}$$

The authors believed that this equation satisfactorily describes the kinetics of PU network formation in IPNs for both stepwise and simultaneous formation of both networks.

The kinetics of simultaneous IPN formation was studied [172] for the system comprising PU prepared with hydroxy-terminated polybutadiene (HTPB), diisocyanate, and chain-extending agent and styrene–DVB or PMMA cross-linked with EGDMA. The authors obtained the curing curves for individual networks and IPNs, and established that the formation rate of the PU component was higher than that for the other two networks under the chosen conditions. The reaction orders for PU and IPN are close to 1, whereas for cross-linked copolymers it is about 0.5. They concluded that the formation of IPNs is dominant.

The search for new ways of IPN synthesis led to a very simple but original solution of performing IPN synthesis according to a radical mechanism in situ [265]. To obtain the IPN, butyl acrylate (BA) or BMA and diallylcarbonate–bisphenol A were used. The originality of the method consists of the choice of such compounds, which initiate the reaction only in a definite temperature range. AIBN was used for polymerization of acrylic monomers at temperatures above 323 K, whereas for allyl monomers *tert*-butylperoxy-3,5,5-trimethyl hexanoate was taken, which initiates the reaction above 370 K. The reaction mixture containing both initiators at the monomer ratio of 50 : 50 by mass was studied by FTIR. The conversion was calculated from the change of the absorption band at 1639 cm^{-1}, which is typical of C $=$ C bonds, with a shoulder at 1649 cm^{-1} corresponding to allylic monomers. At 323 K the intensity of the peak at 1639 cm^{-1} diminished and reached the baseline, corresponding to more than 95% conversion of C $=$ C groups in BMA or BA. By increasing the temperature, the above mentioned shoulder is suddenly transformed into an isolated peak with intensity unchanged up to 370 K. Above this temperature the peak intensity diminishes, indicating the polymerization of the second component. After the introduction of a cross-linking agent, a full IPN can be obtained. The additional IR data confirmed that by using various initiators, but one and the same mechanism of polymerization, semi- and full IPNs can be obtained. Unfortunately, there were no data on the phase state of these systems. Also, from the general considerations it can be assumed that the components of the IPN cannot be compatible.

The kinetic features were studied experimentally and theoretically for IPNs made of PU and unsaturated polyester (PE) [266]. In this case the PU component formed from diisocyanate, polyol, and diol consists of rigid and soft segments. The PE component was synthesized by copolymerization of unsaturated PE with styrene. The grafting of both components is possible through reaction between NCO groups with the end OH or COOH groups of the PE molecule. Such systems can be considered as grafted, which is schematically presented in Scheme 1.

The kinetic features were exposed for two cases of IPN formation: via reaction injection molding (RIM) and via usual molding. Kinetic measurements were performed at isothermal conditions using the DSC method and have shown that due to a different reaction mechanism (polyaddition and polymerization), urethane formation always begins after the mixing of all components. The reaction proceeds under conditions when the polyester component is not yet reacted and urethane formation proceeds in a solution of polyester network components. Polyester network begins to form when PU is either fully or partially formed, i.e., in the medium of the other network. Again, in this work the question about phase separation had not been considered.

The formation of IPNs under adiabatic conditions, which is the model for the RIM process, has shown that the reaction proceeds sequentially and that the dependence of the component conversion on time has an S-shape form, which is typical of radical polymerization. Increasing the amount of PU in the IPN changes the shape of the curve as the urethane formation reaction becomes dominant.

Using additive models [266] based on the additive contribution of IPN components, theoretical calculations were performed for IPNs under consideration of both isothermal and adiabatic regimes. For these calculations, the

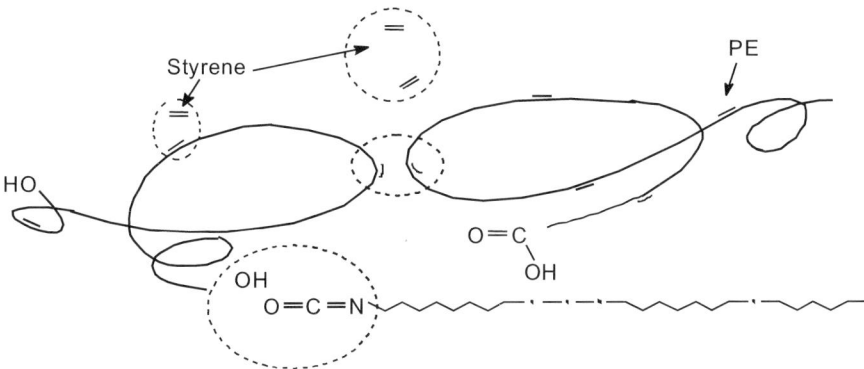

Scheme 1 Schematic structure diagram of PU–PES IPN reactions [266]

Arrhenius equation was used under a set of simplifications that account for no mutual influence of forming network and physical interactions in the system.

The kinetic parameters of the reaction were calculated from the data on the increase of reaction temperature under adiabatic conditions using the following equation:

$$\ln \frac{dT_u}{dt} = \ln \left[\frac{\Delta H_u A_u}{\rho_U C_{PU}} \right] + \frac{E_u}{RT_u} + n \ln \left[\frac{T_{ad} - T_u}{T_{ad} - T_{uo}} \right], \tag{154}$$

where T_u, T_{ad}, and T_{uo} are the reaction temperature, the measured maximum adiabatic temperature, and the initial material temperature, resepctively; ΔH_u is the heat of the PU reaction; ρ_U and C_{PU} are the density and the heat capacity of PU; E_u is the activation energy; A_u is the coefficient of the reaction rate; and n is the reaction order. Variables dT_u/dt, $1/T_u$, and $[(T_{ad} - T_u)/\Delta T_{ad}]$ can be calculated from the dependence of temperature on the time of adiabatic polymerization. Using the method of linear regression, the reaction order n, the activation energy E_u, and the coefficient of reaction rate A_u were calculated.

The kinetic parameters of radical polymerization of polyester were calculated using the Lee method [267] according to the equation:

$$\left(2f I_0 A_p \right) \exp \left(- \frac{E_p}{RT} \right) = \frac{K_d \exp \left[- K_d \left(t_m - t_z \right) \right]}{1 - \exp \left[- K_d \left(t_m - t_z \right) \right]}, \tag{155}$$

where f is the initiator efficiency, I is its concentration, K_d is the decomposition reaction constant, and t is the induction period. The authors reached the conclusion that theoretical calculations for reactions of pure PU and PE agree very well with experimental data obtained under isothermal conditions. For IPN PU/PE with 25:75 mass % compositions essential deviations were observed, with an increase in the value of the mass ratio to 50:50 mass % leading to an increase in the deviations. The authors explained these deviations by the fact that the component interactions were not taken into account. The deviations were much more pronounced for polyester polymerization, possibly due to diffusion effects at high conversion degrees. The mathematical description of the model reaction in the adiabatic regime agrees perfectly with experimental data due to high reaction rates in this regime. It is worth noting that the question about interconnection between the reaction kinetics and microphase separation in the system was not discussed in any of these works.

The kinetics of IPN formation for sequential and simultaneous reactions was studied for IPNs made from cross-linked PU and unsaturated PE [268]. PE was cross-linked by styrene. Typical time dependencies of conversion for both components are given in Fig. 54 for simultaneous IPNs with component ratios 50 : 50 and 30:70 mass %. Although both reactions start at about the same time, the PU reaction is much faster. This is because in externally cat-

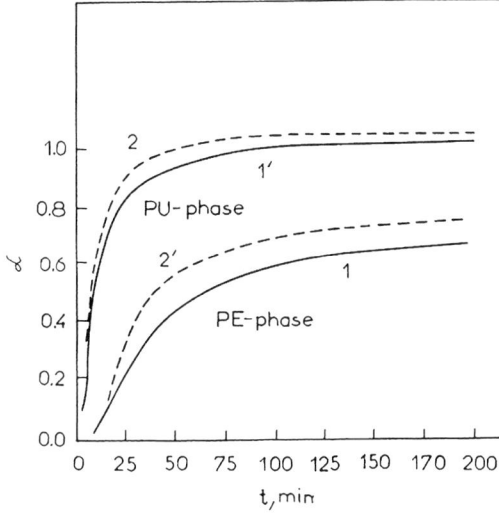

Fig. 54 Dependence of conversion on time for simultaneous IPNs at 333 K for PU (1,2), PE (1′,2′), and various networks of PU/PE ratio: (1,1′), 50 : 50, (2,2′), 30 : 70 by mass % [268]

alyzed step-growth polymerization involving PU, the maximum rate usually occurs at the beginning of the reaction. On the other hand, the free-radical polymerization of PE is still slow and the maximum rate does not appear at the beginning of the reaction. Varying the composition of the IPN changes the reaction rates of PU and PE. Both reaction rates are increased when the mass fraction of PE increases from 50 to 70 mass %. In order to compare the isocyanate conversion at equal stoichiometry, a set of IPNs was prepared by adding extra isocyanate to the mixture to offset the hydroxyl and carboxyl groups in the PE phase. The conversion curves for sequential reactions are given in Fig. 55. Apparently, at equal stoichiometry, the isocyanate consumption rate is lower than that occurring at low NCO/OH ratio. The amount of PE has little effect on the PU reaction at low conversions. But when PE starts to react, there seems to be a slight increase of isocyanate consumption. Both sequential and simultaneous reactions are accelerated by increasing the amount of PE from 50 to 70 mass %. The authors explain this effect for PE by the increasing concentration of double bonds, whereas for PU there is a possibility of interaction between NCO groups of PU with OH groups of both polyol in PU and end OH groups in PE. Therefore, the conversion in the PU phase increases proportionally to the PE concentration in the system. This effect makes the system more complicated, as it implies the possibility of chemical interaction between two components and of the formation of graft networks. In the work under consideration, PU curing proceeds in the solution of initial components that have not yet reacted or reacted only par-

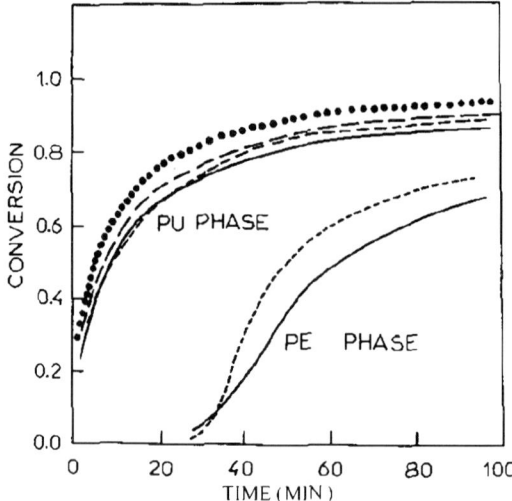

Fig. 55 Conversion vs time for sequential PU/PE IPNs at 333 K: (-) 50 : 50 PU/PE, (· · ·) 30 : 70 PU/PE, (—) 50 : 50 PU/PE at equal stoichiometry, (- - -) 30 : 70 PU/PE at equal stoichiometry [268]

tially. Simultaneously, polymerization of PE proceeds in the presence of the PU network formed earlier. The authors believe that the PU network is already "rigid" when the PE network is formed, and because of this there are two effects on the reaction rates: a "solvent effect" by PU network formation and a "solid effect" by formation of the PE network, both influencing the reaction mechanism.

Within the framework of formal kinetics, the authors derived the equation for the total rate of reaction proceeding via three channels: styrene with styrene, styrene–PE, and PE–PE. The reaction rates of styrene monomer, the conversion of $C = C$ bonds in PE, and the total consumption of double bonds may be described as

$$- \, dc_s / d\tau = k_{ss} \cdot c_{s} \cdot c_s + k_{es} c_{e} \cdot c_s \tag{156}$$

$$- \, dc / d\tau = k_{se} c_s \cdot c_e + k_{ee} \cdot c_{e} \cdot c_e$$

$$- \, dc / d\tau = - \, (dc_s / d\tau + dc_e / d\tau) = k_{ss} \cdot c_s c_s + k_{es} c_{e} \cdot c_s + k_{se} c_s \cdot c_e + k_{ee} \cdot c_e c_{e} \, ,$$

where s denotes styrene, e is $C = C$ bonds in unsaturated PE, s' and e' are radicals of styrene and PE, c is the concentration, and k is the reaction rate constant. The reaction rates were determined experimentally. Special attention was paid to the characteristics of copolymerization between PE and styrene in the PE phase. It was found that in the PU network the PE–PE reaction is more favorable at low conversions as compared with the styrene–PE reaction. This effect was explained by the enhanced intermolecular reaction between PE molecules in the presence of PU. At high conversions, the styrene–PE reaction is more favorable.

In [269] a series of simultaneous IPNs based on PU and PE with different ratios of the components have been studied. The curing process was followed using DSC and FTIR spectroscopy. The curing kinetics of PE and PU is modified greatly during the formation of the IPNs. PE and PU react at a higher rate in the IPNs than in pure homopolymers due to the different mobility of the medium and the existence of collateral reactions. The cross-linking density of the phase that gels first, the conversion reached when this gelling occurs, and the existence of a graft reaction are determining factors in the morphology of the IPN formed together with the existence of phase separation. The sequential IPNs show a lower tendency to phase separation than simultaneous IPNs, as when the PE begins to react both the curing of the PU and graft reactions have been completed. The modules–composition curve follows the Budiansky equation, which predicts phase inversion in the range of intermediate compositions. Three different morphologies with phase inversion near IPN PE/PU 40:60 mass % have been obtained. Compositions that are rich in PU show a continuous PU matrix with disperse PE, while formulations that are rich in PE show the opposite. Intermediate compositions show the existence of two continuous phases.

Simultaneous semi-IPNs composed of epoxy networks based on the diglycidyl ether of bisphenol A (DGEBA) cross-linked with diaminodiphenylmethane (DDM), in which the linear polymer component is polysulfone (PSn) or polyethersulfone (PES), were prepared and their curing process was investigated using DSC and TMDSC [270]. The reaction kinetics was discussed using simple models describing the chemical kinetics including catalytic effects and the influence of diffusion. No significant difference in curing time dependence of T_g (DSC and TMDSC) between the pure network and the DGEBA/DDM-rich phase of the semi-IPNs was found. With decreasing reaction temperature the final glass DGEBA/DDM-rich phase for the semi-IPNs decreases. With increasing PSn or PES content or with decreasing reaction temperature the final conversions were found to decrease, which corresponds to less perfect network structures. Characteristic curing times determined by calorimetry were shown to depend on curing temperature in an Arrhenius-like manner, which is in agreement with dielectric relaxation spectroscopy and mechanical measurements on the same systems.

The polymerization kinetics for IPNs based on vinyl ester resin (VER) and an imidazole-cured epoxy resin (DGEBA) have been studied by DSC and isothermal FTIR. The chemical interactions between the VER initiating system (four different types of radical initiators) and the epoxy curative (two types) have been examined [271]. The AIBN-containing IPNs did not show significant interaction effects between the VER and epoxy component. However, all of the peroxide-initiated IPN systems exhibited an apparent redox reaction between the 1-MeI amine for epoxy component and peroxide, causing an accelerated rate of cure of VER. Dilution effects of the reacting DGEBA system by the VER component were observed for IPNs in the early stages of the cure. During the isothermal cure at 70 °C, unreacted DGEBA monomer plastified the IPN allowing a higher plateau conversion of the vinyl groups in the IPN, provided there were no strong interactions between the 1-MeI and radical initiators. In contrast, when the conversion of the VER component was near complete, the subsequent reaction of the epoxy was limited by vitrification of the IPN associated with the high level of cross-linking in the VER component.

A kinetic study on simultaneous IPN formation of epoxy resins based on DGEBA and unsaturated polyester (UP) was performed by means of DSC. The kinetics of UP and DGEBA reactions was described by empirical models. The DGEBA in a 50:50 mass % UP/DGEBA blend indicated a higher reaction rate constant than the pure DGEBA. The obtained results suggest that the hydroxyl end group of UP in the blend provided a favorable catalytic environment for the DGEBA cure [272].

Simultaneous semi-IPNs composed of a dicyanate resin and PES were prepared, and their curing process was investigated using DSC [231]. The curing rate of the semi-IPN systems decreased as the PES content was increased. The reaction kinetics of the semi-IPNs systems was described by a second-order autocatalytic kinetic equation. The reaction kinetics parameters were determined from the dynamic DSC conversion data by a fitting method. For the semi-IPNs containing 30 mass % PES, two glass transitions indicating phase separation were observed. SEM micrographs showed a phase-separated morphology and different fracture characteristics for the dicyanate/PES semi-IPNs. A very important conclusion as to the effect of molecular topology on the reaction-induced phase separation has been reached in [273]. It has been shown that self-assembled nonreversible morphologies, as well as low-dispersity nucleation and growth, can be controlled through the rate of network formation. The mechanism of phase separation is dictated by the rates of the cross-linking reactions.

The features and detail of the IPN kinetics were also studied in other works [274–276]. The kinetics of thermally initiated cationic epoxy polymerization and free radical acrylate photopolymerization were investigated in [277]. It was found that the preexistence of one polymer has a significant effect on the polymerization of the second monomer. The reaction kinetics and phase separations were studied for sequential IPNs in [278]. The kinetics of IPN formation was studied for IPNs based on PDMS–cellulose acetate butyrate [279]. All these and other works [280–282] confirm the general regularities of the reaction kinetics and its connection with phase separation in forming systems.

5.3
Mutual Influence of the Constituent Networks on the Kinetics of IPN Formation

Preceding data have already shown the influence of the constituent networks on the formation of each network in IPNs. Chemical aspects of the formation of IPNs were discussed in detail by Widmaier and Meyer [283]. They considered the synthesis of in situ sequential PU/PMMA IPNs, including the kinetics of formation of the PU and methacrylic network. The rates of polymerization R_p were deduced from the slopes of the conversion vs time curves. All the parameters influencing the kinetics of the methacrylic system, i.e., PU content, cross-link density, and temperature were studied. The authors constructed the rate vs conversion profiles for all investigated systems.

The influence of the reaction components on the rate profiles was established. The effect of phase separation in the given type of IPN [284] was also taken into account. It was assumed that the formation of small domains allows monomer diffusion toward a radical until its total consumption, and thus phase separation does not hinder the complete monomer conversion. In reaction kinetics a great role was ascribed to the viscosity of the reaction medium. The initial rates increase as the reaction medium becomes more viscous and the onset of the gel effect also begins at smaller conversion degrees. PU behaves like a viscous medium in which the methacrylic polymerization proceeds. Raising temperature shows the known acceleration effect. From the temperature data, the activation energy of the PMMA system was calculated: the values found are about 19 kcal·mol^{-1} whatever the conversion ratio for cross-linked PMMA as well as for all IPNs. This result means that these parameters do not influence the activation process of the methacrylic monomers. The authors conclude that the resulting properties of IPNs are largely influenced by the network formed first and also by the kinetic features leading to the formation of two networks, regardless of the IPN composition.

It is evident that effects of the mutual influence of constituent networks on the kinetics of their formation will be seen more clearly for simultaneous IPNs. This case is also interesting because it allows us to elucidate simultaneously the role of microphase separation in the reaction kinetics.

Studying the kinetics of the simultaneous semi-IPN formation from PU and PBMA [320] has shown that in simultaneously proceeding reactions increasing initiator concentration and rate of BMA polymerization diminish the rate of PU formation as compared to the pure network. It is probable that retardation of PU formation is connected with increasing viscosity of the reaction medium. The higher the rate of PBMA formation, the lower is the rate of PU formation. Indeed, in the case under consideration, the BMA polymerization proceeds in a liquid medium formed by the initial components of the PU network, as if the system were diluted. Because of it, the initial rate of BMA polymerization diminishes as compared with the polymerization rate at the initial stages for pure BMA. However, later, the polymerization rate of BMA in the semi-IPN becomes higher as compared with individual PBMA. The increasing rate may be only connected with the total increase of the viscosity of the reaction system. In accordance with Eq. 152, increasing viscosity leads to sharp diminishing of the termination constant K_t as compared with the growth constant K_g and, correspondingly, to increasing ratio K_t/K_g and chain growth rate V_g. The rate of PBMA formation increases in the sequence $65:35 < 75:25 < 85:15$ as compared with the rate of polymerization of pure BMA. This effect relates to the increasing viscosity. The rate of PU formation diminishes as compared with pure PU in the series $85:15 > 75:25 > 65:35$.

The kinetics of formation of the PU/PS IPN prior to gelation was studied in [286]. It was observed that the extent of PS formation in the IPN increases with increasing PU content, which indicated that the polymerization of styrene was accelerated by the high viscosity of the reaction medium.

Of interest is the study of the kinetics of simultaneous IPN formation from PU and PS using IR spectroscopy [287]. The effects of the temperature, catalyst, initiator, and cross-linking agent concentrations, as well as of the PU/PS ratio, were studied in relation to the rate of formation of both network components. It was established that with increasing PU/PS ratio, the rate of PU formation increases, whereas the PS formation rate diminishes. Changing the catalyst concentration for PU formation has no effect on PS reaction; the same effect was discovered in relation to the initiator concentration.

Authors have found that the rate of the reaction of urethane formation and the rate of radical polymerization of styrene for pure systems are higher as compared with the same reactions by IPN formation. The results obtained for styrene polymerization seem to be rather strange, as both the catalyst and the initiator concentrations in the reaction system were chosen in such a way that the rate of PU formation should be much higher as compared with PS. Therefore, PS was formed when the conversion degree for PU reached 0.4–0.6, and when the viscosity of the medium was rather high. It is known that the termination rate by polymerization sharply diminishes with increasing medium viscosity, and the total rate of polymerization after self-acceleration should be higher as compared with the rate of formation of pure PU. The authors made no attempt to explain the effects of the catalyst or initiator concentration on the formation of two networks. Meanwhile, both processes are tightly interconnected, as was shown in [88, 288].

In [289] the kinetic study was performed for simultaneous IPNs made of PU and PUA. The aim was to establish the effect of simultaneously forming a second network on the rate of reaction of forming the first network. PU based on POPG (MM 2000) and 2,4-TDI adduct was used. The matrix PUA network was formed simultaneously from the oligourethane acrylate (OUA) based on POPG, 2,4-TDI, and the methacrylic ester of ethylene glycol. Figure 56 shows the kinetic curves of PU formation by the simultaneous and uncontrolled reaction of PUA network formation. Since the dependence of $\alpha/(1-\alpha)$ vs t (α is

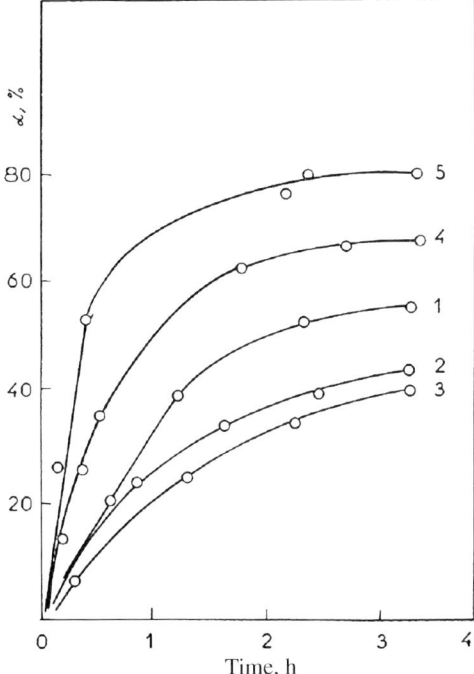

Fig. 56 Kinetic curves for curing of (1) pure PU and PU in an IPN with PUA at (2) 1%, (3) 5%, (4) 50%, and (5) 70% PUA by mass in reaction mixture [88]

conversion degree) is linear up to a conversion degree of 0.5, it may be assumed that the reaction is of the second order. It is worth noting that the initial slope of the kinetic curves depends on the content of OUA in the reaction mixture. By differentiating these curves, the dependence of $\partial\alpha/\partial t$ on the composition has been obtained, and the influence of the component ratio on the reaction rate has been characterized (Fig. 57). The additives in OUA serve as inhibitors of the reaction. For comparable amounts of both components in the mixture, the presence of OUA in the initial stages leads to an increase in the reaction rate, whereas after 45–60 min, the rate of PU formation decreases as compared with the reaction rate in the absence of PUA.

In OUA, both in individual compounds and especially in the mixture, gel formation begins during the first 5–6 min of the reaction. As a result, under conditions of microphase separation the rapidly growing OUA network releases the reacting components of the first network (PU), thus intensifying their self-association. The increase in concentration in separated microvolumes leads to an acceleration of PU formation. PUA evolved as an independent phase may play the role of dispersed filler, which causes an orienting effect on the molecules of the PU components, after which they are adsorbed at the interface. This may also accelerate the process. Inhibition of the pro-

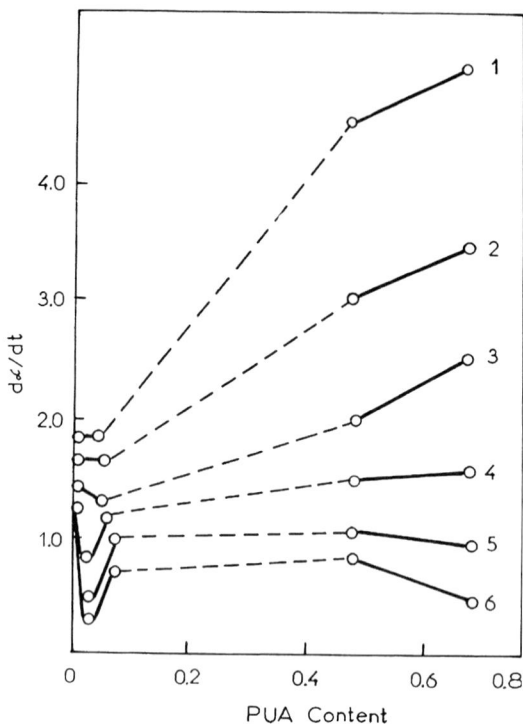

Fig. 57 Concentration dependence of the curing of PUA in an IPN on composition at the time of curing: (1) 10, (2) 20, (3) 30, (4) 45, (5) 60, (6) 90 min [88]

cess at later stages of IPN production is most likely explained by the fact that when the reacting components in microvolumes are exhausted, the reaction is controlled by the diffusion rate of components via the interphase. Thus, it was established that during the synthesis of IPNs the same component, OUA, depending on its fraction in the reaction mixture, might accelerate the degree of completeness of microphase separation. The dependences of the rate and the degree of completeness of microphase separation on the composition of the mixture are behind this phenomenon.

The spectroscopic study has shown that this effect is not connected to chemical interaction between two networks. However, the redistribution of the system of hydrogen bonds takes place in the IPN when the component ratio changes. In this way it was found that the system of intermolecular interactions in a cured IPN, which is controlled by the phase separation in the reaction system, is a function of the composition. IPNs with comparable amounts of constituent networks, as will be shown in detail later, undergo microphase separation, which usually begins before the onset of gelation in the individual networks. This process is characterized by a high rate and degree of segregation (separation).

A more detailed study of the kinetics of IPN synthesis has shown that the close relation between the kinetics and phase separation really exists [288]. For IPNs made from PU and PEA the kinetic curves of cross-linking were obtained. The rate of PU curing was affected by the ratio of the TMP adduct with TDI and oligoglycol and the rate of PEA curing by the initiator concentration (0.01% in series 1, 0.5% in series 2, and 3.0% in series 3). Figures 58 and 59 [289] show kinetic curves for the curing of PU and PEA networks taken both separately and in a mixture for IPN formation. Judging from the slopes of the curves, the cross-linking rates of each network in the IPN are not independent. This effect is not associated with chemical interaction between networks. By differentiating the kinetic curves presented in Figs. 58 and 59 it is possible to determine the curing rate $\partial \alpha / \partial t$ of PU and PEA at each point of the curve. The dependence of $\partial \alpha / \partial t$ on the composition of IPNs at different concentrations of an initiator, for a fixed time since the initiation of the reaction, gives a series of curves characterizing the influence of the constituent network on the rate of PU (Fig. 58) and PEA (Fig. 59) curing at different stages of IPN formation. The reaction rate of urethane formation in IPNs, especially at the initial stages, depends both on the concentration of oligoesteracrylate (OEA) and on its curing rate (Fig. 60). In the region of low amounts, the PEA network inhibits PU cross-linking. The degree of inhibition is maximal in the initial stages of reactions and depends on the ratio of the curing rates of individual networks. OEA most effectively inhibits PU cross-linking in the samples of series 1 where it is cured at a lower rate than the individual PU.

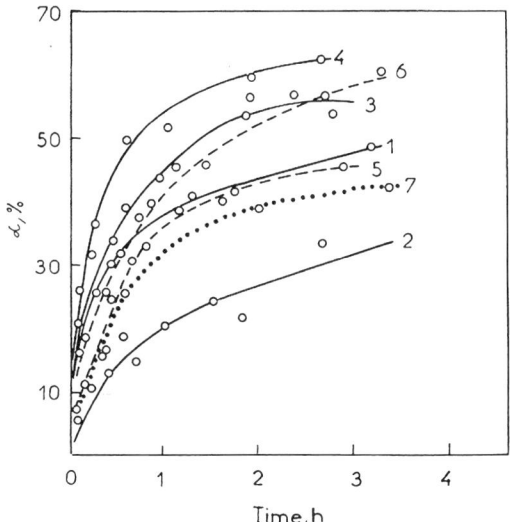

Fig. 58 Kinetic curves of pure PU (1) and of PU in IPNs with 0.3 (5), 0.5 (2), 2 (3, 6, 7), and 20 (4) by mass % of PEA. Curves 1–4: series 1; curves 5, 6: series 2; curve 7: series 3 [289]

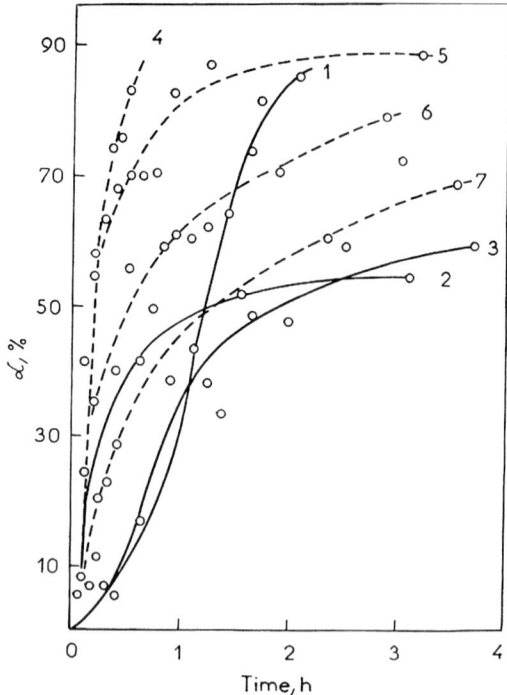

Fig. 59 Kinetic curves of pure PEA (1,4) and of PEA in IPNs with 10 (5), 30 (2, 5), and 50 (3, 7) mass % PU. Curves 1–3 refer to series 1, curves 4–7 to series 2 [289]

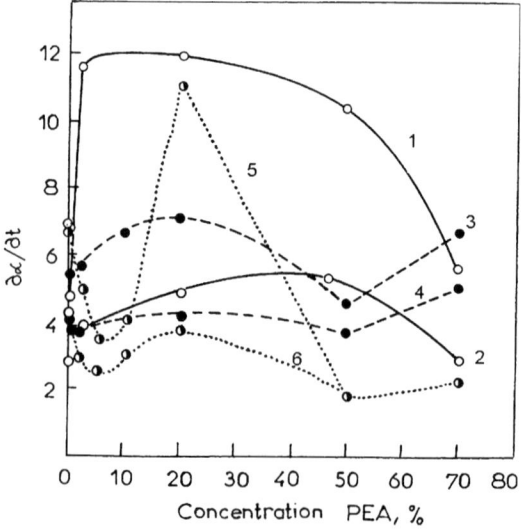

Fig. 60 Dependence of the rate of PEA curing on the composition of IPNs through 5 (1, 4), 10 (2, 5), 20 (6), 40 (3), and 50 (7) min after the reaction initiation. Curves 1–3 refer to IPNs with 0.01% of initiator, curves 4–7 to those with 0.5% of initiator [289]

Five minutes after the reaction is initiated, the rate of urethane formation in IPNs of this series is 1.5 times lower than that for individual PU. In series 2, OUA, which is cured at a higher rate, is inhibited less effectively than the process of matrix network cross-linking. In series 3, retardation of the urethane formation reaction is observed over a wide range of compositions.

In the region of average compositions, OEA accelerates PU cross-linking as early as the initial stage of the reaction. As in the case of inhibition, the effect depends on the rate of OEA curing. The PEA network initiates the curing of PU most effectively with 0.01 and 3.0% concentration of initiator.

The influence of OEA on PU network formation weakens with the development of a three-dimensional structure, with the rate of the adduct cross-linking in IPNs approaching the curing rate of pure PU. The PEA network accelerates PU curing at later stages of the reaction in IPNs with low addition of OEA. At the same time, the rate of PEA formation in the IPN depends not only on the initiator amount but also on the PU content (Fig. 61). In this case the influence of the latter compound depends on the curing rate of the OEA. In series 1, PU, which is cured at a higher rate, initiates OEA cross-linking

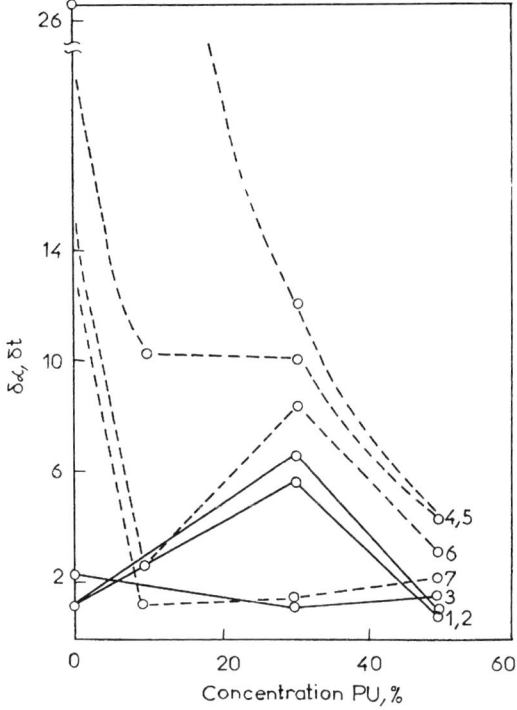

Fig. 61 Dependence of the rate of PEA curing on the composition of IPNs through 5 (1, 4), 10 (2, 5), 20 (6), 40 (3), and 50 (7) min after initiation. Curves 1–3 refer to IPNs with 0.01% of initiator, curves 4–7 to those with 0.5% [289]

in the range of composition studied. At the same time, in series 2, where the curing rate is lower than that of OEA, PU inhibits the process.

With the beginning of microphase separation, the influence of the second network on PU curing decreases and the curing rate of PU in the IPN approaches the cross-linking rate of pure PU during reaction. Analysis of the system spectra recorded during the reaction demonstrates that an increase of OEA concentration leads to a decrease of the time of microphase separation (i.e., the time when the bands of self-association of urethane groups connected with formation of the microregion and "rigid" blocks begin to appear). The release of the PEA network under microphase separation evidently removes the PU network components, thus intensifying their self-association, which accelerates the urethane formation reaction. In addition, the appearance of microphase regions of PEA may be due to the adsorption interaction of PU components with PEA, which accelerates its curing. In other cases, a sharp increase of viscosity prevents the diffusion of components and inhibits microphase separation. Thus, the rates of constituent network curing in IPNs are connected with each other through the microphase separation process.

5.4
Relation Between Reaction Kinetics and Microphase Separation

The above analysis of the kinetics of constituent network formation in IPNs shows that these processes are connected and accompanied by microphase separation of the system. The reasons for such an effect consist of the appearance of the fragment immiscibility of growing networks.

5.4.1
Semi-IPN Based on Styrene–DVB Copolymer and PBMA

Figure 62 shows the anamorphoses of kinetic curves for the formation of semi-IPNs from styrene–DVB copolymer in the presence of varying amounts of PBMA [88, 290]. The absence of the inflection point on the curves at the time when phase separation begins (conversion degree 0.1–0.8) is typical. The gelation in the system begins only after 60–65% conversion. The lack of inflection point may be explained by the argument that the phase separation proceeds according to the spinodal mechanism of decomposition, when at the initial stages the compositions of the evolved phase are very similar. Increasing the amount of PBMA in the reaction mixture decreases the time of the onset of phase separation. Increasing the temperature by 30 °C also diminishes the onset of phase separation, although the conversion degree of styrene is increased. This result was the first where it was shown that there exists an interconnection between the reaction kinetics, conversion degree, and microphase separation in the system, with the kinetics of IPN formation

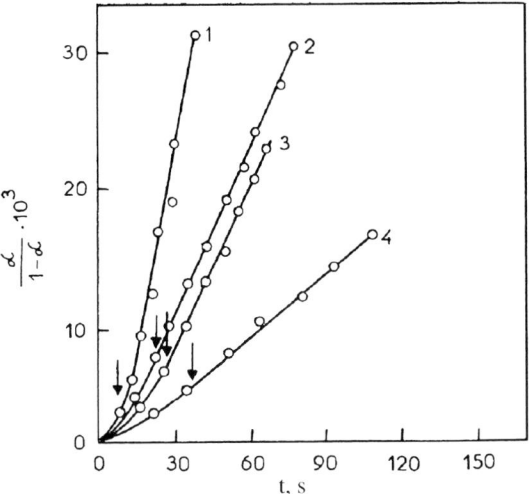

Fig. 62 Kinetic curves for copolymerization of styrene with DVB in semi-IPNs at 363 K: (1) 40%; (2) 15%; (3) 40%; (4) 15% PBMA by mass. *Arrows* show the onset of phase separation [88]

exerting a marked effect on the microphase separation, determining the onset of phase separation [88, 290]. The higher the degree of conversion, the lower is the fraction of the second component, at which the phase separation proceeds (Table 12) [88].

As can be seen from this table, at equal degrees of conversion, the time of phase separation onset is higher at lower reaction temperatures. At these conditions, the compatibility of components is lower. However, the viscosity of the system is high enough to prevent the separation. As a result, the onset of phase separation is detected experimentally at higher conversion degrees. Therefore, one can conclude that the chemical reaction and the phase separation proceed simultaneously. Depending on the reaction rate, cross-linking

Table 12 Times of the onset of phase separation (t) and corresponding conversion degrees (α) in semi-IPNs at various temperatures [290]

PBMA, mass %	333 K α	t, s	343 K α	t, s	353 K α	t, s	363 K α	t, s
40	0.12	2280	0.90	1140	1.93	840	0	60
20	0.73	5400	1.55	2880	2.00	1200	2.68	700
15	1.45	7200	2.28	3060	2.44	1500	2.88	900
10	1.90	10920	2.40	5460	4.46	2280	2.91	1080
5	4.75	21600	5.40	9900	6.27	3600	7.56	2160

significantly lowers the molecular mobility of reaction components. Thus, phase separation can be stopped at various degrees of conversion. Therefore, by changing the kinetic conditions of the chemical reaction, one can influence the structure and the properties of IPNs. Phase separation, proceeding simultaneously with the cross-linking reaction, occurs under nonequilibrium conditions. These conditions are determined by the relations between the reaction kinetics, the conversion degree, and the kinetics of phase separation. Here, of great importance is the mechanism of phase separation, i.e., nucleation and growth or spinodal decomposition. The gelation time depends also on the amount of the second component; the activation energy is dependent on this factor as well (Figs. 63 and 64).

The detailed investigation of this process was done using two types of semi-IPNs made of cross-linked PU and linear PBMA [202]. The difference between these two cases is that in one case the previously polymerized PBMA was introduced into the reaction mixture for PU network formation, whereas in the other case the reactions of PU formation and BMA polymerization proceeded simultaneously. PU based on POPG and TDI–TMP adduct was formed under the action of dibutyltin laurate (catalyst). The kinetics of PU network formation and BMA polymerization were studied by calorimetry. The kinetic parameters of the reaction for pure PU and for PU in the presence of the second component were found from equation of the second-order reaction:

$$K = \left(\frac{\alpha}{1 - \alpha}\right) \frac{1}{A_0 t}, \tag{157}$$

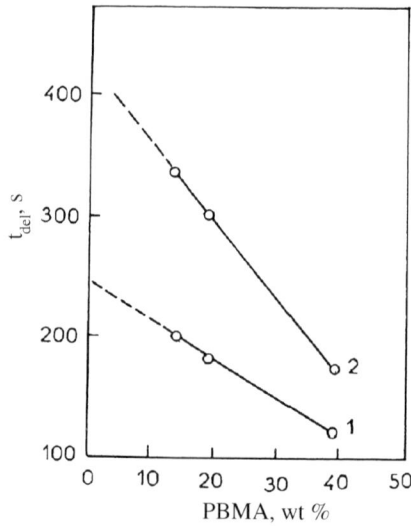

Fig. 63 Dependence of the gel formation time in semi-IPNs on the PBMA content: (1) 363 K, (2) 353 K [88]

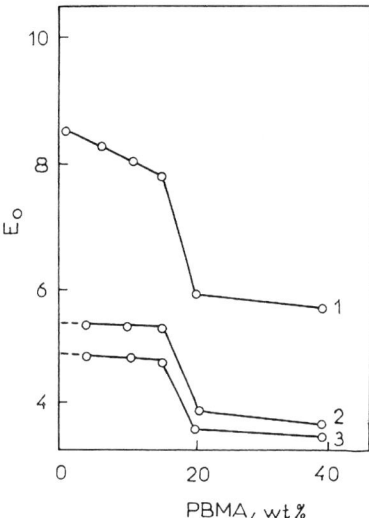

Fig. 64 Dependence of the apparent activation energy on PBMA concentration: (1) initial stages of the reaction, (2, 3) latest stages of phase separation, (2) according to dilatometric data, (3) from the time of the onset of phase separation [88]

where α is the conversion degree, A_0 is the initial concentration of reacting groups, and t is the reaction time. The reaction rate constants for urethane were calculated from the linear dependence of $\alpha/(1 - \alpha)$ on t. To establish the interconnection between kinetics and phase separation, the segregation degrees were determined from viscoelastic properties. These values are a measure of incomplete phase separation in the systems with restricted thermodynamic compatibility. The onset of phase separation in the system was followed by the cloud point method.

5.4.2
Semi-IPN Formed by Curing PU Network in the Presence of PBMA

Figure 65 presents anamorphoses of kinetic curves of the formation reaction of cross-linked PU in the absence (curves 1, 2 at 333 K and 3 at 313 K) and in the presence of PBMA (curves 4–6 at 333 K, 7 at 353 K, and 8 at 313 K), investigated at various temperatures and catalyst concentrations. Both for pure PU and for semi-IPNs the obtained anamorphoses exhibit two parts, which evidence a change of the reaction rate in three systems on reaching a certain degree of conversion. It can be noted that a comparison between experiments 1 and 4, 2 and 5, or 3 and 8 shows a threefold decrease of the urethane formation rate in the presence of PBMA, while the character of anamorphoses of the kinetic curves remains qualitatively unchanged. Analyzing these results, it should be noted that only in one case (curve 8) does the degree of con-

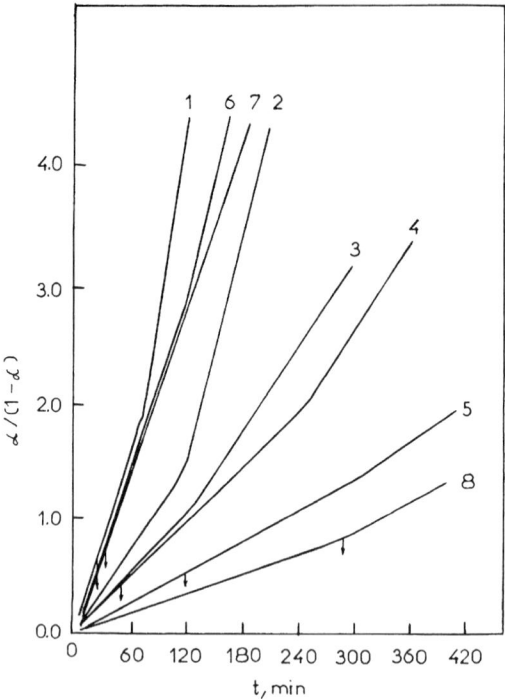

Fig. 65 Anamorphoses of kinetic curves of the initial PU formation process (1–3) and in the presence of 15 mass % PBMA (4–8) at various catalyst concentrations $[kt] \cdot 10^4$, mol·l^{-1}: 0.7 (1, 5); 1.4 (2–4, 7, 8); 7.0 (6) and temperatures, °C: 40 (3, 8); 60 (1, 2, 4–6); 80 (7). *Arrows* show the beginning of phase separation [202]

version at which microphase separation begins coincide with the conversion corresponding to the inflection of anamorphoses. It is suggested [289, 291] that at low degrees of conversion, because of the higher viscosity of the system, microphase separation slows down and gets experimentally observed only at higher degrees of conversion, which in fact corresponds to the appearance of the thermodynamic incompatibility of the components. The absence of inflections on the kinetic curves at the microphase separation onset is most likely due to the absence of volume changes brought about by microphase separation, since this separation itself does not affect the rates of reactions. The obtained data do not indicate a change in reaction kinetics at the start of microphase separation near the degree of conversion corresponding to it. The above-presented result demonstrates that in the system studied, the onset of microphase separation depends substantially on the urethane formation kinetics.

5.4.3
IPNs Produced by Simultaneous Curing
of Polyurethane and Polymerization of Butyl Methacrylate

For further study of the features of IPN formation, kinetic investigations were made [202] when formation of the PU network and polymerization of BMA and formation of PBMA proceed simultaneously and in a stepwise manner. Kinetic curves for the initial PU and PBMA with different amounts of initiator are shown in Fig. 66. As can be seen, the PBMA formation rate (curves 9–11) increases with initiator concentration. Using the principle of additivity and determining the PU conversion degree by an independent method, the heat release area corresponding to PBMA formation, and hence the degree of BMA conversion into a semi-IPN, was established from the difference between the total heat release area and the area corresponding to PU formation, calculated from its conversion degree. Kinetic curves for PU and PBMA in a semi-IPN at PU/PBMA ratio 75:25 mass % and molar concentration of initiator of 1.48×10^{-2} mol·l^{-1} are given in Fig. 66 (curves 13a and 13b). It was shown that for simultaneously proceeding reactions, the rate of BMA polymerization increases with initiator amount (Fig. 67, curves 13b, 14b, 15b, 16b) while the PU network formation rate declines with respect to the initial one (Fig. 67, curves 13a, 14a, 15a, 16a). As the PBMA formation rate increases, the degree of conversion rises at a certain moment, which increases the viscosity of the medium and thereby impedes diffusion of macromolecules that form PU chains (Fig. 67).

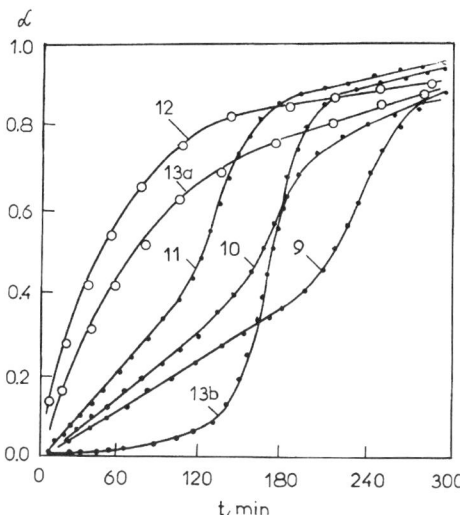

Fig. 66 Kinetics of PU formation and BMA polymerization at various initiator concentrations [I]·10^2, mol·l^{-1} for BMA: (9) 0.74, (10) 1.48, (11) 2.96, (12) PU, (13a) PU in semi-IPN (75 : 25 by mass %), and (13b) BMA in semi-IPN [202]

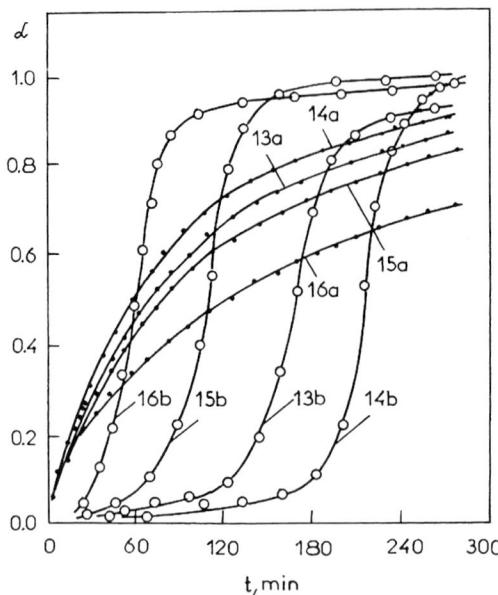

Fig. 67 Kinetics of PU formation (a) and BMA polymerization (b) in semi-IPNs at various initiator concentrations, $[I] \cdot 10^2$, mol·l^{-1}: (14) 0.74; (13) 1.48; (15) 2.96; (16) 1.08 [202]

The comparison of the reduced rates of PBMA formation, $W_{red} = V/M$, where $V = dM/dt$ is the amount of unreacted monomer and M is monomer concentration, in the semi-IPN and the initial rate (Fig. 68) shows an increase in W_{red} of PBMA formation in the mixture with respect to the initial one (Fig. 68, curves 11 and 15b, 10 and 13, 9 and 14). This, according to [311], may be due to an increase of the initial viscosity of the semi-IPN system, which rises in the course of PU and PBMA formation, thereby sharply reducing the termination constant K_t and accordingly increasing the K_g/K_t ratio and the chain growth rate V_g with increasing completeness of reaction, according to Eq. 152.

Figure 69 presents kinetic curves for PU and PBMA for various ratios in the mixture: 85 : 15, 75 : 25, and 65:35 mass % at a constant molar concentration of the catalyst (1.4×10^{-4} mol·l^{-1}) and various initiator concentrations. Calculated W_{red} values for PBMA in the mixture show that the PBMA formation rate increases for experiment 19 to experiment 17 over the initial one (curves 19a, 18a, 17a), which is due to the increase in the system viscosity; in experiment 17 it is higher than in experiment 19, and hence K_t decreases, while K_g and K increase. The PU formation rate in the mixture decreases compared to the initial one from experiment 17 to experiment 19 (curves 19a, 18a, 17a).

A change in the rate of formation of both PBMA and PU is thus observed in a polymeric system where two processes proceed concurrently. It should

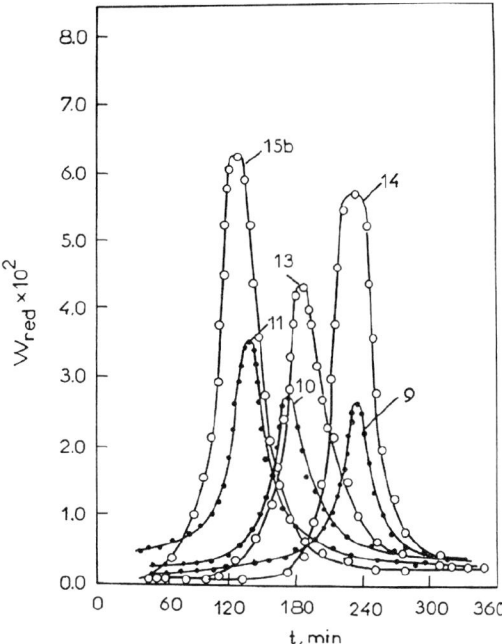

Fig. 68 Variation of reduced rate W_{red} of BMA polymerization at various initiator concentrations $[I] \cdot 10^2$, mol·l^{-1}; (11) pure BMA, $[I] = 2.96$; (10) pure BMA, $[I] = 1.48$; (9) pure BMA, $[I] = 0.074$; (15b) BMA in IPN, $[I] = 2.96$; (13) 1.48, (14) 0.74. Ratio PU/PBMA 75 : 25 by mass % [202]

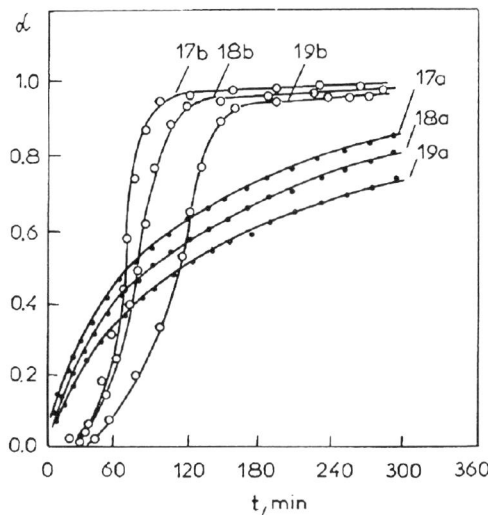

Fig. 69 Kinetics of PU formation and BMA polymerization in semi-IPNs at various component ratios: a—PU in IPN, b—PBMA in IPN. PU/PBMA: 17–85/15, 18–75/25, 19–65/35 by mass % [202]

be pointed out that the PU formation rate, regardless of the ratios of components and initiator, drops as compared to the initial rate, whereas the PBMA formation rate depends on any factor increasing the system viscosity, which is associated with the gel effect. It can also be noted that an inverse dependence of the PU formation rate on the PBMA formation rate is observed in a series with different molar concentrations of initiator. This is a consequence of various conditions of PU and PBMA formation in the mixture.

The analysis of the data shows that the time of the microphase separation onset t_{mps} decreases almost linearly with increasing concentration of the initiator of BMA polymerization. Calculations of the BMA polymerization rate at the initial stage (until the autoacceleration starts) demonstrate linear dependence between the reaction rate and t_{mps}. The data show that at constant initial composition of the system, PU conversion drops and BMA conversion increases at the moment of the onset of microphase separation with increasing polymerization rate. The higher the BMA polymerization rate, the greater its conversion drops at the onset of separation and the smaller the PU fraction at which separation begins. Correspondingly, t_{mps} decreases.

Microphase separation also depends on the PU/PBMA ratio. Indeed, for the ratio 85:15 mass % (Table 13) the conversion degree before the onset of microphase separation is higher for both components than for the ratio 75 : 25, while the values for the ratios 65 : 35 and 75 : 25 are close to each other. At the same time the PU mass fraction corresponding to t_{mps} turned out to be the

Table 13 Compositions and microphase separation parameters for PU–PBMA semi-IPNs [285]

PU/PBMA ratio, mass %	$[I] \cdot 10^2$, mol·l^{-1}	PBMA rate constant $\cdot 10^3$, s^{-1}	Onset of separation time, min	Conversion degree at the separation onset, PU	Conversion degree at the separation onset, PBMA
75 : 25	0.74	0.70	70	0.57	0.034
75 : 25	1.48	0.83	65	0.50	0.040
75 : 25	2.96	2.66	55	0.38	0.062
75 : 25	5.40	3.80	45	0.36	0.080
75 : 25	10.80	6.60	35	0.21	0.090
65 : 35	5.40	2.20	65	0.37	0.110
85 : 15	5.40	5.5	60	0.45	0.280

lowest for 65 : 35 and the highest for 85 : 15. It should be taken into account that the microphase separation rate in a high-viscosity medium should be distinctly lower than the reaction rates, and therefore microphase separation may lag in time behind the conversion at which it would occur if it proceeded under equilibrium conditions, not aggravated by the continuing reaction.

Studies of the influence of the microphase separation on the IPN formation kinetics show [330] that at a constant component ratio, the PU conversion at t_{mps} drops and BMA conversion rises with increasing PBMA formation rate, the influence of the initiator concentration being more pronounced than that of the cross-linking BMA agent (triethylene glycol dimethacrylate) concentration. Such regularity was traced for two component ratios, 75 : 25 and 65:35 mass %. The value of t_{mps} in the system drops sharply only with a component ratio of 65 : 35 and at higher initiator concentration. When the component ratio is changed at constant initiator and cross-linking agent concentrations, t_{mps} increases to 70 min, the PU conversion declines from 0.5 to 0.3, and the BMA conversion varies within 0.03–0.20 [292].

It follows that the polymerization rate growth, increasing the BMA conversion, shortens t_{mps}. The most essential inference from the analysis of reaction rates and values of t_{mps} is, however, the following. As can be seen from the kinetic curves, both before and after microphase separation their character remains unchanged, i.e., microphase separation does not affect the rates of continuing reactions. This implies that in the evolved microregions of phase separation, the composition becomes close to the average composition of the system, i.e., the ratio of reacted and unreacted components is approximately the same. As was already suggested, such a situation can occur only with microphase separation at the initial stages proceeding through the spinodal mechanism. This obviously involves "chemical quenching" of the microheterogeneous system resulting from network formation, which after a certain conversion prevents further phase separation according to Binder and Frisch [93]. Otherwise, the microphase separation would progress and departures of the reaction rates from their values in the initially separating system should occur, which is not observed right up to a high degree of conversion.

5.4.4
Thermodynamics and Kinetics of Phase Separation

The effect of various kinetic conditions on the phase separation in semi-IPNs produced from PU and PS was studied using FTIR by He, Widmaier, and Meyer [293]. To obtain semi-IPNs of PU/PS mass ratio 30 : 70, the mixture of initial components was stored at room temperature for different periods of time t (5 min, 15 min, and 2 h), and then the temperature was elevated up to 343 K and the process was performed until the full disappearance of the double bonds.

It was discovered that at $t = 5$ and 15 min, PU network and PS are formed simultaneously but with different rates, whereas at $t = 2$ h the formation of PS proceeds in a fully cured PU network. Various kinetic conditions of PU formation affect the phase separation estimated from the data of the light transmission as a function of time. The system loses its transparency by simultaneous formation of PU and PS ($t = 5$ min) when the conversion degree of PS is about 0.2. By forming the semi-IPN in the $t = 2$ h regime the system stays transparent, whereas for $t = 15$ min a trend toward diminishing transmission of light is observed at the latest stages of IPN formation.

Introduction of preliminary polymerized PS into PU in an amount of 0.5–1.0% initiates phase separation in the system in the first few minutes. It was also found that increasing MM of PS diminishes the time of the onset of phase separation. Thus, the authors reach a conclusion that if the reaction mixture is separated into two phases before gelation of PU, the semi-IPNs formed are turbid, but if separation begins after PU gelation transparent IPNs are formed. The thermodynamic reasons for these effects are not clear.

It can therefore be inferred that simultaneously proceeding chemical processes of formation of polymer molecules which constitute the IPNs and physical processes of microphase separation occur under nonequilibrium conditions [103]. In this case the microphase structure of semi-IPNs and the kinetics of their formation become interrelated (see Sect. 4). The effect of the reaction conditions on the morphology of simultaneous PU/PMMA IPNs has also been established in [294], and for poly(dimethylsiloxane–urethane)/PMMA IPNs in [295]. It was shown that depending on the kinetic conditions, both compatible and incompatible IPNs could be formed with well-matched rates of cross-linking reaction. Incompatible IPNs are formed by a much slower cross-linking reaction producing phase-separated IPNs. We believe that in the present case, because of the method of determination, one should talk not about true compatibility, but about the dependence of apparent compatibility on reaction kinetics.

The first attempt to describe theoretically the processes of phase separation during the reaction of formation of semi-IPNs has been done in the works [296, 297]. Semi-IPNs based on PS and a reactive epoxy monomer based on DGEBA with a stoichiometric amount of 4,4′-methylenebis(2,6-diethylaniline) were studied experimentally. Thermodynamic analysis of the phase separation proceeding during the curing reaction was performed that considered the composition dependence of the interaction parameter $\chi(T, \Phi_2)$ (where T is the temperature and Φ_2 is the volume fraction of PS) and the polydispersity of both polymers. The latter is especially important. In this analysis, $\chi(T, \Phi_2)$ was considered as the product of two functions, one depending on the temperature $[D(T)]$ and the other depending on the composition $[B(\Phi)]$. For the initial mixture (before the reaction) the cloud point curves showed upper critical solution temperature behavior and the dependence $\chi(T, \Phi_2)$ on the composition was determined from the threshold point, that is, the maximum cloud point temperature. During the isothermal reaction of mixtures with different initial PS concentrations, the dependence of $\chi(T, \Phi_2)$ on the composition was determined under the assumption that at each conversion degree, the $D(T)$ contribution to the $\chi(T, \Phi_2)$ value had to be constant independently of composition. For these semi-IPNs it was demonstrated that the changes in the chemical structure produced by the reaction reduced $\chi(T, \Phi_2)$. This effect was more important at lower volume fractions of PS. Nevertheless, the decrease in the absolute value of the entropic contribution to the free energy of mixing was the principal driving forced behind the phase separation process.

The theoretical model developed by the authors was proved by the good agreement between experimental and predicted glass transition temperatures and heat capacity changes at the T_g for both evolved phases. For every degree of conversion, the theoretical simulation of the phase separation provided the amount, composition, and conversion of each phase. From their analysis, they have reached the following conclusions. At the beginning of the polymerization, the system is homogeneous. As the reaction proceeds, the system becomes less miscible and at a certain conversion level (the cloud-point conversion) the system becomes phase separated. For the system studied the authors consider the following possible scheme of phase separation. At the cloud-point conversion, the composition of the continuous α phase is the initial one, and the volume fraction of the β dispersed phase is zero. The segregated phase has a large amount of the epoxy–amine polymer, and conversion in this phase is higher than that of the continuous phase as a result of the polymer fractionation due to the entropic effect. After the cloud point it was thought that the epoxy-rich phase was semipermeable. This signifies that the β phase will receive material from the continuous phase. All the effects strongly depend on the process chemistry. However, the kinetics of chemical transformations at this reaction stage was not studied. Finally, the segregation is almost complete and the mass fraction of the continuous phase is close to that of the initial composition. Unfortunately, the authors do not pay any attention to the mechanism of phase separation (spinodal mechanism or nucleation and growth). The only note is that for compositions near the critical composition, the reaction proceeds to drive the system into the unstable region of the phase diagram and then the phase separation proceeds according to the spinodal mechanism. For off-critical compositions (the case considered by the authors) the phase separation proceeds by nucleation. They have not considered the degree of the thermodynamic equilibrium of the cured system and do not take into account the very probable formation of an interfacial region between evolved phases, which may be formed easily in reaction systems possessing high viscosity (incompleteness of phase separation). Thus, thermodynamic analysis and experimental data confirm the qualitative picture of the process of phase separation proceeding by the formation of IPNs proposed earlier in our works.

5.5
Reaction Kinetics and Properties of Evolved Phases

The kinetic effects influence not only the onset of the phase separation but also the composition and the ratio of evolved phases as well. It was shown that it is possible to evaluate the ratio of phases and their composition from the shifts of glass transition temperatures of each phase with respect to the pure component prevailing in this phase. Because of this, the elucidation of the kinetics of the transitions in evolved phases is of great importance for understanding of the interrelation between the reaction conditions and phase separation proceeding in the systems.

The kinetics of the formation of these systems was studied [202] for the PU/PBMA IPNs. As distinct from the preceding results, in this work the phase separation resulting from kinetic changes was estimated by the glass transition temperatures of two evolved phases, which characterize their composition and hence the conditions of phase separation. Table 14 shows some parameters of semi-IPN PU/PBMA 85:15 mass % obtained by introduction of linear PBMA into the reaction mixture. The main criterion of phase separation in this work is the shift of glass transition temperature of the two evolved phases, each of which is enriched by one of the components.

The glass transition temperatures of the evolved phases depend on the way PBMA was introduced: I is the commercial PBMA introduced into

Table 14 Parameters of PU/PBMA IPNs (85:15 mass %) [297]

$[kt] \cdot 10^4$, mol·l^{-1}	Glass transition (K) of the phase enriched in PU	PBMA	Time of the onset of microphase separation, min	Segregation degree
0.7	261	333	100	0.32
1.4	263	316	43	0.33
7.0	248	323	28	0.35

Table 15 Parameters of semi-IPN PU/PBMA (75:25 mass %) [297]

$[I] \cdot 10^2$, mol·l^{-1}	Transition temperature of the phase enriched in PU	PBMA	Segregation degree, α
0.74	265	323	0.11
1.48	263	333	0.12
2.96	263	330	0.28
5.40	263	358	0.28
10.8	258	351	0.28

the reaction mixture (Table 14), II is PBMA formed from monomer in situ (Table 15).

As is seen in most cases, the glass transition temperatures do not coincide with the corresponding temperatures of pure components. This is a result of the formation of a two-phase system with phase composition determined by the reaction conditions (which in their turn determine the conditions of phase separation). Usually, as a result of formation of two phases enriched in one of the components, glass transitions approaching each other are observed. Such a situation is really observed only for IPNs with PBMA introduced into the reaction mixture beforehand. In all other cases the T_g of the PU-enriched phase either does not change as compared with the pure network or becomes lower (Table 15). This effect may be explained [202] by the formation of a more defective network structure of PU itself due to concurrence of two reactions proceeding simultaneously. It can be suggested that the microphase separation, which begins at the initial stages of reaction, affects the network structure of PU. The effect of the reaction rate determines the properties of the phases evolved. However, independent of the way PBMA was introduced, the PU network becomes more defective. The decrease in the rate of PU network formation due to changing catalyst concentration really leads to the formation of a distinct structure of the network, as can be judged from the value of T_g for the PU-enriched phase. Introduction of monomer and its polymerization has the same effect. Increasing polymerization rate decreases the T_g of the PU-enriched phase, whereas the segregation degree increases. If the cross-linking agent is introduced into the reaction system T_g practically does not change. These data evidence that the kinetic conditions of reaction affect the composition of phases evolved during reaction.

Microphase separation in semi-IPNs based on epoxy resin and linear polymers (PSn and PES) formed by isothermal curing was studied in [270] using usual and TMDSC. It was found that in the presence of linear polymers the curing rate diminishes owing to a decrease in diffusion coefficients and the density of reactive groups. The final conversion diminishes with increasing amount of linear polymer and with decreasing curing temperature, which leads to a less perfect network structure. The authors studied the time evolution of the two-phase system from measurements of the glass transition temperatures. It was shown that at the beginning of the reaction, the system is characterized by one glass transition temperature (one-phase system). With developing reaction the second glass transition temperature appears. This temperature changes with time up to a constant value. The intensity of the glass transition steps, which is characterized by the heat capacity increment by transition, depends on the composition of the evolved phase.

Thus, the most typical feature of the kinetics of IPN formation is a superposition of chemical and physical processes, which distinguishes these systems from all other polymeric materials. The analysis of the data concerning the effect of the reaction rate on microphase separation allows the following conclusions about the phase state of simultaneous IPNs to be drawn:

1. If one of the networks is formed much faster than the other, its formation proceeds in the liquid medium of the second network components. In this case the phase separation may be rather complete and be frustrated. The evolution of the swollen phase of the first network in the liquid medium is possible. The second network is formed later and, as a result, a system with rather high segregation degree arises.
2. Both reactions of formation of two networks proceed with high rates. Microphase separation has no time to proceed and the structure is frozen, which is typical for a one-phase state [295]. The system is not in equilibrium and is characterized by the "quenched" structure of the initial or intermediate reaction mixture. This case should be typical for IPNs produced by the RIM process.
3. The most common case—both networks have similar reaction rates. The microphase separation begins in the course of reaction, its development is hindered by the formation of the network structure, and it stops at a certain degree of cross-linking (or viscosity growth). In this case the two-phase structure is determined by the ratio of the reaction rates of the two constituent networks. The phase composition also depends on the reaction rates and on the degree of microphase separation at the point of ceasing separation due to cross-linking. When a linear polymer is introduced in the case of semi-IPNs, the rate of network formation has the same effect on the degree of microphase separation.

A special case of the formation of IPNs is RIM [298]. IPNs comprising a cross-linked acrylic (Ac) polymer as one component and a polyurea (Pur), segmented copolyurea (co-Pur), or copoly(urea–isocyanurate) (co-PurI) as second component have been formed by the RIM process. The effects on the processability and formation of the IPNs of the cross-linker concentration in the acrylic component, the functionality of the amine-functionalized polyether used for the polyurea, and the mass fractions of acrylic components were evaluated. The reaction kinetics during RIM processing of the IPNs was studied using adiabatic temperature rise (ATR) measurements. IPN formation in Ac/Pur and Ac/co-PurI occurred sequentially: in Ac/Pur IPNs, the first-formed polymer was the rubbery poly(ether–urea) component, whereas that in Ac/co-PurI was the cross-linked acrylic component. The use of a triamine-functionalized polyether in Ac/Pur IPNs resulted in almost instantaneous chemical gelation of the Pur component, which caused very poor mixing with the acrylic-forming component and produced very heterogeneous materials. In contrast, the use of the diamine-functionalized polyether allowed efficient mixing and produced homogeneous materials. Decreasing the degree of cross-linking in the Ac component in Ac/Pur IPNs, by increasing the mass ratio of MMA/diacryl, improved the mixing efficiency during processing and resulted in more complete polymerization of the Ac component. The use of a higher concentration of initiator for the Ac component and catalyst for the co-PurI component increased the overall rates of formation of the IPNs. The degree of transparency of the IPN materials depended on the relative proportions and domain sizes of the components formed during RIM processing, consistent with the differences observed in ATR reaction kinetics.

The results discussed above meet the concept of the nonequilibrium state of IPNs. In all cases the phases evolved have the composition corresponding to the state of mixing at the earlier stages of the reaction, because the microphase separation is hindered due to the viscosity effect and cross-linking.

5.6
Kinetics of Formation of Sequential IPNs

Reactions of formation of sequential IPNs have their specifics, because after curing the first component of the IPN the polymerization (or polycondensation) of the second polymer proceeds in the matrix (guest) of the first polymer. In [299] the kinetics of simultaneous and sequential IPNs based on PU and PMMA have been studied by FTIR methods. Simultaneous IPNs were prepared at 333 K and in situ sequential IPNs were made at room temperature (PU network completely formed first) and at 333 K (PMMA network formed second). It was established that sequential IPNs have a higher degree of phase dispersion than the simultaneous IPNs by using the methods of dynamic mechanical analysis (DMA), TEM, and NMR spectroscopy. The authors tried to explain these results by the fact that in the former type of IPN, the PU host network is completely formed before the onset of copolymerization of the second component: its presence precludes the formation of large rigid domains, contrary to the case where the two networks formed simultaneously possess a morphology with larger domains.

In [300], blends of varying composition of a bisphenol A based cyanate ester (BACY) and a bisphenol A based bismaleimide (BMP) were cured together in a sequential manner to derive bismaleimide–triazine polymer networks. In the presence of DBTDL as catalyst, the BACY–BMP blend underwent sequential polymerization, each component polymerizing independent of the other. The cure characterization of the blends was done by DSC. The cured polymers were characterized by FTIR and thermogravimetric analysis (TGA). The cured blends were found to undergo two-stage decomposition, each stage corresponding to the polycyanurate and polybismaleimide. The high-temperature stability was substantially improved on incorporation of a potential network interlinker containing both maleimide and cyanate functions.

Sequential IPNs based on PMA and PMMA were prepared by UV photopolymerization [301]. The PMA/PMMA system is immiscible and so, for low cross-link densities, phase separation appears, as detected by the occurrence of two clearly differentiated main dynamic mechanical relaxation processes corresponding to the two components. The IPNs cross-linked with 10 mass % EGDMA show a single main dynamic mechanical relaxation process. Since the length of cooperativity in the glass transition region has been determined to be around a few nanometers, this fact suggests that any region of the IPN (interfacial region) with this approximate size contains both components. Permanent physical internetwork entanglements are also characteristic of the compatibilization. It has been shown that the key factor controlling the miscibility in the IPN is the cross-linking density of the network polymerized first. Very interesting detailed results were presented in [302, 303]. The authors have investigated the kinetics of PS and PMMA formation on previously prepared networks based on copolymers of MMA with triethylene glycol dimethacrylate and styrene with EGDMA by photopolymerization. It was established that polymerization of styrene and MMA from the very beginning starts from the stage of the gel effect. The growth constants by polymerization of these monomers in the networks coincide with those by polymerization in bulk, whereas the termination constants are lower. The latter effect leads to the earliest appearance of the gel effect. The molecular mass distribution of PS and PMMA was determined after alkaline hydrolysis of the obtained semi-IPNs. The data on the value of M_w/M_n for PS equal to 3.0 show that under conditions preventing chain termination, the polymer formed has less polydispersity as compared with the same polymer synthesized in bulk ($M_w/M_n = 4.1$).

The dependence of the phase separation process on the relative onset of network formation in simultaneous IPNs was studied in [304]. The morphology development during synthesis at room temperature of a PU/PMMA IPN was investigated by SAXS in relation to their relative kinetics of formation, determined by FTIR spectroscopy. When the time lag between the onsets of the two reactions is short, macroscopic phase separation occurs as the PU network is incompletely formed. However, when the time lag increases, PMMA forms into a more continuous network, which limits the growth of phase separation to a close environment.

A series of in situ sequential IPNs of PU and PS was prepared in [305]. The PU network was made from POPG, end linked with an aliphatic triisocyanate. The PS network results from free radical photopolymerization of styrene with small amounts of DVB. During synthesis, the homogeneous initial mixtures segregate into co-continuous phases where the interpenetration takes place. The number of additional entanglements between unlike chains increases with increasing cross-link density of the primary PU network. For the same system both polymerizations were performed either simultaneously or one after the other [306]. It was discovered that the degree of phase separation depends on the experimental conditions. Two levels of phase separation can be distinguished in IPNs made according to the in situ "simultaneous" process:

(a) a macroscopic phase separation into almost pure phases, resulting from the competition between the rate of network formation and the rate of phase separation during the early stages of synthesis, i.e., before gelation of the reaction mixture; and (b) a microphase separation thermodynamically controlled by the inherent level of miscibility of constituents. The results suggest that controlling the chemistry and process (cross-linking density, composition, and time sequence of events) of in situ IPN formation will give various morphologies and hence properties, ranging from microphase-separated materials to larger macrophase-separated materials.

5.7
Role of Kinetics in Formation of an Interphase and Segregation

The reaction kinetics of IPN formation plays a very important role in the formation of the final structure. It determines both the segregation degree and fraction of the interfacial regions. In some of our works for semi- and full IPNs the effect of kinetic parameters on the microphase structure has been investigated in detail. For IPNs based on cross-linked PU and styrene, BMA, and MMA, the effect of the following factors on the process has been investigated [307–313]: the ratio of components in the starting reaction mixture,

Table 16 Parameters of microphase separation for PU/PS IPNs at various concentrations of catalyst and initiator, and various PU/PS ratios [312]

Semi-IPN PU/PS (70 : 30) at $[I] = 1.0 \times 10^{-2}$ mol·l^{-1} and various concentrations of catalyst

$[kt] \cdot 10^5$, mol·l^{-1}	Reaction rate constant, $K \cdot 10^5$, kg·mol^{-1} s^{-1}	Time of the onset of phase separation, min	Segregation degree α	Fraction of interphase $(1 - F)$
0	5.0	45	0.20	0.20
0.3	10.2	18	0.31	0.18

Semi-IPN PU/PS (70 : 30) at $[kt] = 0.3 \times 10^{-5}$ mol·l^{-1} and various initiator concentrations

$[kt] \cdot 10^2$, mol·l^{-1}	Reaction rate constant, $K \cdot 10^5$, kg·mol^{-1} s^{-1}	Time of the onset of phase separation, min	Segregation degree α	Fraction of interphase $(1 - F)$
1	10.2	18	0.31	0.18
5	8.0	25	0.21	0.26

Semi-IPN at various PU/PS ratios, $[kt] = 0.3 \times 10^{-5}$ mol·l^{-1} and $[I] = 1.0 \times 10^{-2}$ mol·l^{-1}

PU/PS ratio	Reaction rate constant, $K \cdot 10^5$, kg·mol^{-1} s^{-1}	Time of the onset of phase separation, min	Segregation degree α	Fraction of interphase $(1 - F)$
70 : 30	10.2	18	0.31	0.18
50 : 50	8.7	35	0.33	0.26
30 : 70	7.2	50	0.35	0.30

the concentrations of catalyst and initiator of radical polymerization, the time of the onset of microphase separation (by light scattering), the conversion degrees of both components at the onset of phase separation, the segregation degrees α, and the fraction of the interfacial layer $(1 - F)$ in products of the reaction. Some data are given in Table 16.

It is seen that the segregation degree and fraction of the interfacial region depend concurrently on many factors, the conversion at the point of the onset of phase separation being the most important. The segregation degree α is determined by:

1. The rate of the onset of phase separation, i.e., the rate of reaching the critical molecular masses at which the incompatibility arises.
2. The conversion degree at which phase separation begins.
3. The reaction rates of the formation of both components.
4. The MM of both components at the onset of phase separation.

   The point is that the reaction rates of the formation of both components in the reaction mixture depend on the presence of the second component and are not equal to the reaction rates for pure components. In this case one network is formed earlier, and the reaction of the formation of the second proceeds in the matrix of the first. At a higher rate of formation of one of the networks, the critical MM for the onset of phase separation is reached faster.

   It is worth noting that the concentration of the cross-linking agent plays almost no role in segregation because the latter proceeds before reaching the gel point. In a series of special investigations we have studied the in situ formation of blends of two linear polymers, which are formed according to the various mechanisms (the principle of IPN formation). The observed regularities of this process are similar to that for IPNs, thus testifying to the similar mechanism of phase separation in the absence and in the presence of the cross-linking agent.

   Molecular masses at the onset of phase separation are determinative factors in the appearance of the thermodynamic incompatibility of the network fragments. At the same time, up to now there are no experimental data on the MM of both components at various stages of reaction and at the onset of phase separation. For the blends of linear PU and PMMA synthesized in situ, we have determined the MM of both components after completion of the reaction. These values were also estimated for pure components obtained under the same conditions. It was found that the kinetic conditions of the reaction determine the MM and MM distribution by simultaneous curing, which are different from the results of polymerization of pure components. These data testify to the marked effect of the reaction conditions on the MM, and in such a way contribute to the understanding of the effect of reaction conditions on the phase segregation and fraction of an interfacial region.

5. Microphase separation depends also on the ratio of components in the starting reaction mixture. This ratio determines the reaction kinetics and time of the onset of phase separation.

Due to a great number of factors it is not possible to establish a direct relation between them and the segregation degree. However, the general rule is valid: the higher is the segregation degree α, the less is the fraction of an intermediate region $(1 - F)$, which is natural, as the latter is a measure of the unseparated part of the system.

The incomplete phase separation or phase separation due to the spinodal mechanism leads to the formation of a transition zone or an interphase between two evolved phases. This is that part of the system which for kinetic reasons stays in the unseparated state and preserves the structure of the reaction mixture before the onset of phase separation.

5.8
Description of Reaction Kinetics in Terms of Phase Transition

Processes of the network formation may be described within the framework of the phase transition. Gelation proceeds under conditions of the new phase formation and can be described by the Avrami–Erofeev equation [314, 315]:

$$\alpha = 1 - \exp^{-Kt^n} , \tag{158}$$

where α is the conversion degree, t is the reaction time, and K is a constant. This equation is widely applied to describe phase transitions (crystallization, reactions in the solid phase, etc.). The chemical reaction after the appearance of microgel particles proceeds under microheterogeneous conditions and is accompanied by exhaustion of reactive groups both inside the microgel particles and between them, i.e., with participation of the surface groups [316]. In IPNs the reaction also proceeds inside the evolved phases and at the interface between them, although this process has never been investigated. For such a case the process may be described by the Avrami–Erofeev equation with coefficient $K = 1$, i.e., by an equation of phase transition. Representing the bimolecular reaction of cross-linking in terms of the Avrami–Erofeev equation seems to be more preferable than in terms of the second-order reaction. This approach was developed for microheterogeneous systems, in which various microregions exist where the reaction proceeds. For such a case the equation may be presented as [316]:

$$\alpha = 1 - \exp\left(-\sum K_i \tau^{n_i}\right) , \tag{159}$$

where K_i is the specific reaction rate in the i-th zone, and n_i is a parameter characterizing the influence of the i-th type structure on the reaction rate in the i-th zone. It is assumed the changes of n between 1 and 3 characterize the relative contribution of homogeneous and heterogeneous constituents of the curing process. Increasing n implies the localization of the reaction on the surface of the microphase or gel particles. From this point of view it is important to analyze the kinetics of IPN formation in a similar way, keeping in mind not only the formation of microgel particles of the PU network, but also the appearance of microphase particles connected with microphase separation [317]. When the reaction of the IPN begins, it proceeds very quickly under conditions of the microphase-separated system. Increasing conversion degree leads to increasing immiscibility of IPN components and to formation of two phases, in which composition changes continuously in the course of reaction.

To establish the influence of the interface formation on the reaction kinetics of IPN formation, a comparative investigation has been performed [318] of the reaction kinetics of pure PU network formation and the formation of simultaneous semi-IPNs from cross-linked PU based on POPG, TDI adduct, and BMA polymerization (75:25 mass %).

Figure 70 shows the typical dependence of the conversion degree of PU in semi-IPNs in terms of the Avrami–Erofeev equation. The values of K and n in

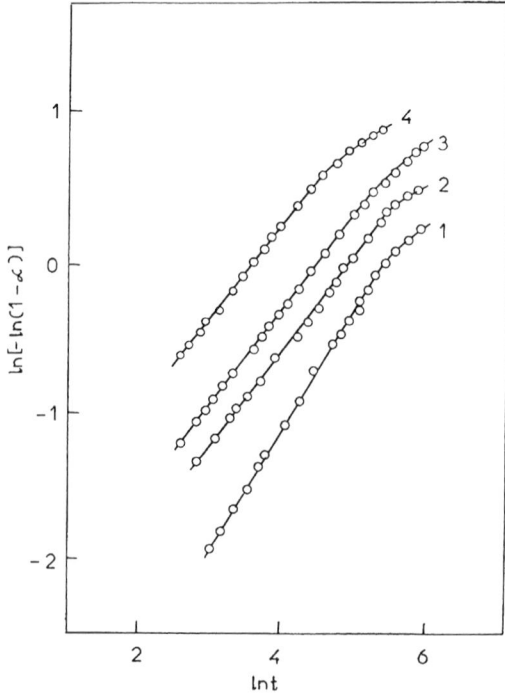

Fig. 70 Dependence of the conversion on time in coordinates of the Avrami–Erofeev equation during PU network formation in IPNs. MM of polyether: (1) 2000; (2) 1500; (3) 1000; (4) 500 [318]

Table 17 Values of the Avrami–Erofeev equation parameters K and n for PU and semi-IPNs at curing temperature 333 K [318]

System	MM of polydiol	n	$K \cdot 10^2$, min^{-1}
PU	2000	0.94	1.10
PU	1500	0.90	1.82
PU	1000	0.91	2.35
PU	500	1.18	3.00
semi-IPN	2000	0.80	1.25
semi-IPN	1500	0.71	3.60
semi-IPN	1000	0.70	5.50
semi-IPN	500	0.65	10.00

this equation are given in Table 17. The value of n for semi-IPNs is lower than for pure PU and depends on the MM of the POPG taken for PU formation. The value of K increases with decreasing MM of polyglycol. Thus, curing of PU in semi-IPNs depends on the presence of PBMA. This effect was explained in the following way. The formation of microgel particles in the reaction system proceeds simultaneously with microphase separation, and the curing of PU proceeds simultaneously inside the particles and on their surface. The curing of pure PU proceeds under homogeneous conditions up to the point where microgel particles are formed (conversion degree $\alpha = 0.17$). Curing PU in the IPN proceeds under nonhomogeneous conditions from early stages of the reaction due to microphase separation. Microphase separation promotes aggregation of microgel particles. Their chemical interaction contributes to the total reaction rate and affects the K value. In this case the process really should be described by a modified Avrami–Erofeev equation (Eq. 159), since this process proceeds both inside and on the surface of particles, K reflecting the contribution of both processes. Because of the formation of the microphase separation regions, the total number of microphase particles in the system increases. Simultaneously, in the zone of microregion formation due to the spinodal mechanism of separation, the microgel particles are formed. The growth of their total surface leads to increased reaction rate on the particle surface and therefore increased K. Parameter n diminishes and becomes less than 1, probably being connected to the formation of microparticles of various shapes (spherical microgel particles and treelike spinodal structures). The process can be represented by the scheme in Fig. 71. As a basis, the scheme of spinodal decomposition was taken from [55]. Double formation of microparticles, according to this scheme, results in the increased area of the interface. Reactions at the interface have a higher rate and higher value of K.

It is worth noting that the Avrami–Erofeev equation is applied to many physical processes and allows varying interpretations of the equation con-

a)   b)   c)

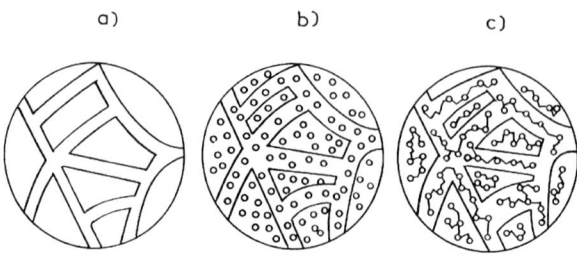

Fig. 71 Schematic illustration of spinodal decomposition at various reaction stages: **a** initial stage, **b** developed stage with the appearance of microgel particles, and **c** late stages and cross-linking of microgel particles in two evolved phases [318]

stants. Because of this, one can conclude that the microphase separation accompanying chemical reaction leads to the change of the shapes of microgel and microphase particles in such a way that this change enhances their interaction. Increasing the value of n, according to [316, 317], means that the main contribution to the reaction belongs to the localization of the cross-linking at the interface of microphase particles. However, we believe that the consideration of the kinetic effects within the framework of the proposed model allows us to explain the processes of cross-linking in IPNs after two phases have already evolved, to explain cross-linking at the interface between phases.

5.9
Some Features of Rheokinetics and Structure States of Semi-IPNs During Their Formation

The formation of IPNs during chemical reactions of curing is accompanied not only by phase separation at a certain conversion degree, but also by other structural and physical changes in the system, which are also determined by the kinetic conditions. In forming IPNs, a great role is played by the diffusion processes, which in their turn depend on the rheological properties and on changes of the molecular mobilities of components in the course of reaction.

The kinetics of curing IPNs made from polybutadiene-based PU [172] and PS-*co*-DVB or PMMA-*co*-EGDMA was studied from measurements of the time dependence of torque. The kinetic parameters of the reaction were calculated and curing curves were constructed. However, in this work the method of investigation was used only for purely kinetic purposes without analyzing the structural and rheological changes in the system during curing.

The change of rheological properties of reaction systems is determined by two main factors: kinetic features and rheological properties of initial and forming components. To establish the conditions of gel formation during reaction, both the kinetics of reaction and the rheokinetics were studied [119, 319]. Simultaneous semi-IPNs produced from cross-linked PU and PBMA formed by polymerization simultaneously with PU formation were used. The ratio PU/PBMA was taken to be 75 : 25 by mass. Figure 72 shows the dependence of viscosity $\log \eta$ on reaction time for PU formation. Two break points are observed: one at a conversion degree of PU $\alpha = 0.17$ and the other at $\alpha = 0.49$. The first break is determined by the formation of a branched prepolymer and by the appearance of the microgel particles. The second break may be ascribed to the development of microheterogeneity in the system due to the onset of structural gelation. Analysis of the rheokinetic curve of PBMA formation shows that the increase in viscosity up to the first break (Fig. 72, curve 2) is connected with the growth in PBMA concentration in a polymerizing system at constant MM. After the break, a sharp increase in reaction rate was observed together with a sharp increase in viscosity. It is known [320] that the autoacceleration during radical polymerization is determined by a dramatic increase of radical concentration. As a result the conversion degree increases, being accompanied by an increase in viscosity. These data were used as a basis for studying the rheokinetics of simultaneously formed semi-IPNs made from PU and PBMA. Figure 72 (curves 3–6) shows rheological curves for the formation of the semi-IPN of composi-

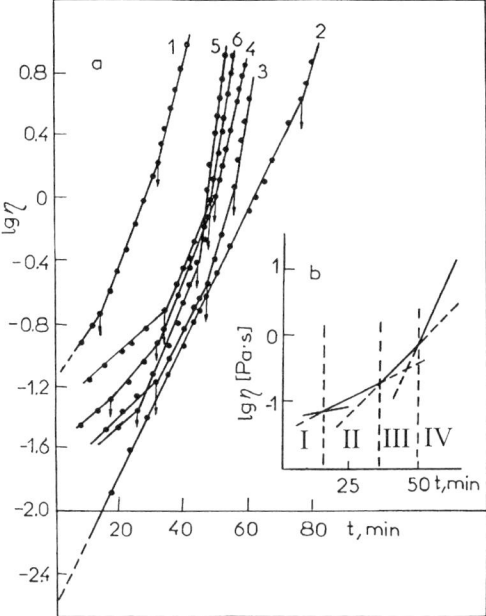

Fig. 72 **a** Dependence of the largest Newtonian viscosity on time during the formation of cross-linked PU (1), PBMA (2), and semi-IPN (75:25 mass %) at initiator concentration 0.74×10^{-2} mol·l^{-1} (3); 1.48×10^{-2} mol·l^{-1} (4); 2.96×10^{-2} mol·l^{-1} (5), and 5.4×10^{-2} mol·l^{-1} (6). **b** Schematic presentation of the same curves. *Arrows* indicate the time of probe selection [119]

tion 75:25 mass % for various reaction rates of BMA polymerization. All the curves are similar and are characterized by three break points at characteristic conversion degrees, their values being dependent on the reaction rates of formation of both PU and PBMA. It may be suggested that the first break reflects the change in viscosity due to PU formation, the second corresponds to the onset of phase separation in the system, and the third is connected with autoacceleration of BMA polymerization in the reaction mixture. Thus, the rheological curves show changes in viscosity with various effects operating, including the initial stages of phase separation, gel particle formation in PU, and finally the gel effect in PBMA. However, it was not possible to ascribe the appearance of any inflection point to a specific chemical or physical process. Most probably various effects, as estimated via rheological analysis, overlap. The break points on the rheological curves do not correspond to the break points for separately cured components. In the opposite case, it should imply that both reactions proceed independently and are not accompanied by the phase separation of the system. Therefore, the deviations in the positions of break points for semi-IPNs from those for individual components are determined by the influence of the reaction kinetics on the rheological behavior and by the phase separation. This is the principal feature of the changes in rheological properties of semi-IPNs as compared to traditional networks. However, the question about structural changes taking place during IPN formation remains open.

Some information can be obtained from the viscoelastic behavior of the curing system [321]. To understand the transformations of the system during curing, the viscoelastic characteristics were monitored [321], and specimens of the IPNs were obtained by stopping the reaction by sharply decreasing the temperature down to 263 K at a certain time, which corresponds to the position of the break on the rheokinetic curves (Fig. 72a). By lowering the temperature, the radical polymerization of BMA breaks off as the initiator acts only above 323 K; at the same time the reaction of urethane formation is also very slow at low temperatures. The real and imaginary parts of the complex shear modulus, the storage modulus G' and the loss modulus G'', were measured at various frequencies. Figure 72 separately shows the rheokinetic curves for the synthesis of IPNs. Arrows mark the times at which specimens were taken. This curve shows clearly the various time regions of IPN formation (denoted by the Roman numerals).

From the rheokinetic curves the dependencies of the reaction time corresponding to the break points on the initiator concentration have been plotted (Fig. 73). Roman numerals denote the same regions of the IPN as in Fig. 72b. To judge the physical state of the system under consideration (initial PU and IPN) the viscoelastic properties have been studied for all the time regions. The dependencies of the real part G' and of the imaginary part G'' of the complex shear modulus on frequency obey the scaling laws $G = \omega^{\alpha}$ and $G'' \sim \omega^{\beta}$. Here the exponents α and β at $\log \omega < 0.5$ are equal corre-

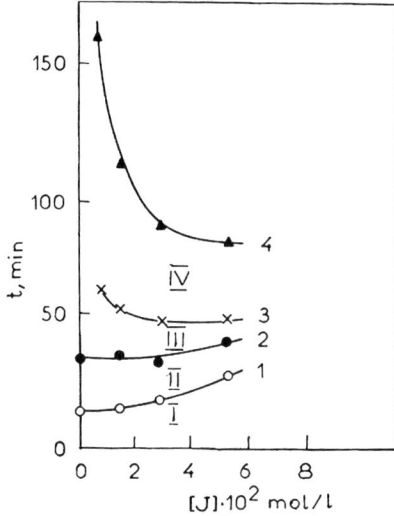

Fig. 73 Dependence of the time corresponding to the break points on the rheokinetic curves for semi-IPNs and PBMA on initiator concentration: curves 1–3 correspond to three regions of the IPN; curve 4 is the same for PBMA [119]

spondingly to 2 and 1 for samples taken after 40 min of reaction. These values correspond to the theoretical ones for flexible polymer chains [322]. At $\log \omega > 0.5$ both exponents decrease in value. Such behavior is typical of linear or branched polymers with a MM above the critical one for entanglement. In such a way, for given kinetic conditions (333 K and $t = 40$ min), i.e., after the second break on the time dependence of $\log \eta$ (Fig. 72a, curve 1), we are already dealing with the developed structure of PU. It is worth noting that when the duration of the reaction of PU curing is less than 15 min (the region before the first break point), the elastic reaction on the deformation in this system is absent (the angle between the stress and the deformation by harmonic oscillation is 90°). For these conditions only the dissipative part of the complex modulus G'' may be reliably registered. For the intermediate reaction times, between 15 and 40 min, the system reveals its viscoelastic properties. The absolute values of G' and G'' increase with increasing reaction time.

Figure 74 represents the frequency dependencies of G' and G'' at the initiator concentrations of 0.74 (left) and 1.48×10^{-2} mol·l^{-1} (right). The same dependencies were obtained for all the concentrations used. This kind of dependence has been extensively discussed in the literature [244].

It is seen that the dependencies $G'(\omega)$ and $G''(\omega)$ also agree with theoretical laws $G' \sim (\omega)^2$ and $G'' \sim (\omega)$ for the end zone of viscoelasticity. For all investigated IPNs, the absence of the elastic part of shear modulus at the early stages of the formation (25–30 min) is typical. As a rule, such a situation takes place

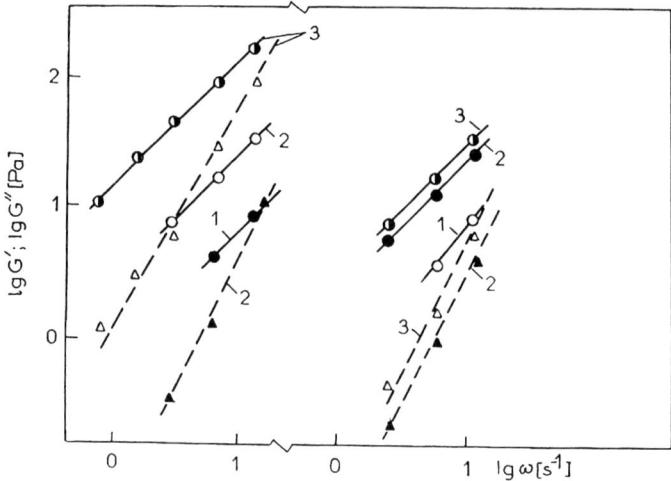

Fig. 74 Frequency dependence of moduli G' and G'' (*solid line*) for semi-IPNs (75:25 mass%) at initiator concentrations of *left* 0.74×10^{-2} mol·l^{-1} and *right* 1.48×10^2 mol·l^{-1}. The duration of IPN formation: (1) 30, (2) 45, (3) 55 min [119]

before the second break on the viscosity curves. The increasing rate of PBMA formation due to an increased amount of initiator also increases the absolute value of G' and G'' at constant ω and t.

The experimental data allow some conclusions to be drawn as to the features of the structural state of the curing system at various stages of the process. Consider the data given in Figs. 73 and 74. We assume that region I (Fig. 73) corresponds to the state of the component solution at the reaction temperature. In the course of reaction the formation of microgel particles of branched PU prepolymer proceeds in the medium of monomeric BMA. The concentration of these particles is rather low and the system is a pure viscous nonelastic liquid. This state corresponds to region II (Fig. 72). Region III is characterized by the appearance in the system of some microheterogeneity as a result of the onset of microphase separation and structure gel formation in PU with simultaneous formation of PBMA molecules (conversion degree does not exceed 0.1). Such a system is a viscoelastic liquid (the viscoelastic part of complex shear modulus is not zero). After the third break on the rheokinetic curve (region IV), the cross-linked structure of PU is already formed, and a gel effect by formation of PBMA is observed. It is possible that this state already consists of two separated phases, although the method used does not allow us to distinguish between the causes of the rheological changes. This state is achieved by the system in a period of time, which depends on the initiator concentration. This can be seen from the dependence of the onset of autoacceleration of PBMA on the initiator concentration (Fig. 73, curve 4). Increasing the total reaction time of IPN formation (region IV) leads to increased network density and as a result, the viscoelasticity appears at lower frequencies (Fig. 74).

Taking into account the change in character of the dependence of t, corresponding to the break points, on initiator concentration during the transition from curves 1 and 2 to curves 3 and 4 (Fig. 73) we can conclude that the change in the law of the non-Newtonian viscosity growth with reaction time by transition from the first to the second break points is connected both to the formation of the PU network and the growth of its fraction in this system. The change of the exponent after the third break point should be ascribed to the gel effect.

These results show that the evolution of the system in the course of its formation seem to be important from the point of view of understanding the real kinetic conditions of IPN formation and phase separation. Rheokinetic data allow us to study various structural states (from a rheological point of view) in different time periods of the reaction. Of course, the discussion given here is over-simplified because it is difficult to separate pure kinetic effects that influence the viscosity and viscoelastic behavior from effects connected to simultaneously proceeding microphase separation.

The data discussed above allow us to conclude that in order to study the features of IPN formation, it is insufficient to have only kinetic data and data on phase separation during reaction. Of great importance are the data on the molecular mobility [95, 323] of reaction components at various reaction stages, and data on the general changes of the system viscosity at various stages of IPN formation.

5.10
Special Features of Self-Organization During the Formation of IPNs

The data presented above permit us to consider special features of IPN formation and self-organization during this process [324]. According to definition [325], self-organization involves the appearance, development, and disappearance of macroscopic structures under nonequilibrium conditions. Self-organization means the appearance and the development of the structure in the initially homogeneous environment. As was shown, the chemical reactions of IPN formation and phase separation in the course of reaction proceed simultaneously and under nonequilibrium conditions.

The data presented here allow it to be established that the following two processes underlie the self-organization during the formation of IPNs: (1) the cross-linking and formation of a three-dimensional network, i.e., gel formation, or sol–gel transition for each network; and (2) phase separation in the system caused by the appearance of thermodynamic incompatibility (immiscibility) of constituent fragments of various networks at a certain degree of conversion. The experimental data make it possible to qualitatively determine the general conditions of self-organization during the formation of IPNs occurring under thermodynamically nonequilibrium conditions.

The initial system for the synthesis of IPNs represents a homogeneous mixture of components, which are able to form two independent networks in the course of chemical reactions proceeding according to different mechanisms. When a certain degree of conversion is achieved (as a rule, long before the sol–gel transition), thermodynamic incompatibility of the propagating fragments of the constituent networks arises in the system. The conditions for its appearance is the attainment of the critical value of the thermodynamic interaction parameter χ_{23} predicted by the theory. The initial conditions of phase separation are the transition of the parameter χ_{23} from a negative value (the compatibility region) to a positive one (the incompatibility region), or the point when χ_{23} at the given composition of the system and the given temperature become zero. In most cases this phenomenon occurs at the initial stages of reaction in its kinetic region. However, phase separation takes place under nonequilibrium conditions from the very beginning, since it proceeds simultaneously with the developing chemical reaction, which changes parameter χ_{23}. The rate at which χ_{23} approaches the critical value is determined by the rates of two independent reactions, and the phase separation process begins before gel formation in the constituent network has been completed.

In the initial period of the reaction (small degrees of conversion) the beginning of phase separation is determined only by the reaction rate, i.e., by the rate of polymer formation and by the rate of increase of its MM. However, cross-linking of the propagating chains into the continuous network and rapid growth have a hindering influence on phase separation, which, in this way, becomes possible only within definite ranges of conversion and time interval Δt, which is a function of the rates of two reactions. In this case, since cross-linking continues within this time interval, the phase separation process occurs under nonequilibrium conditions because the composition of the phases and their ratio changes with time. Thus, the real structure arising as a result of the formation of the IPN differs significantly from the case of "chemical quenching", where it was assumed that the curing time is much shorter than the relaxation time of the system during phase separation. Both processes (chemical reaction and phase separation) proceed simultaneously, making a very interesting example of self-organization. This process determines the character of those structures which develop and transform at different stages of the reaction and phase separation.

It has already been discussed that the phase separation in IPNs may be described in terms of the spinodal decomposition. At the initial stages the kinetics of phase separation is described by the linear theory of Cahn–Hilliard in spite of the fact that continuous changes of conversion occur in this case. This means that the system drifts along the phase diagram with a continuous change in parameter χ_{23} and in the composition of the separating phases. Thus, one more principal feature of IPN structure formation is established. This feature consists in a continuous sequence of structures or states, which differ in composition not only at various stages of phase separation, but also in the character of chemical reaction. Therefore, the microphase structure arising during the reaction is the result of a nonequilibrium phase transition of the liquid–liquid type, and the final result of such a transition is determined by the composition of the system (by the ratio of networks, the reaction rates, and the depth of nonequilibrium phase separation for the time interval Δt).

Thus, general conditions of self-organization during the creation of IPNs may be formulated in a qualitative form. These conditions are determined:

(i) By the ratio of the rates of chemical reactions for the constituent network formation

(ii) By the sol–gel transition and the phase separation depending on which process begins earlier

(iii) By the kinetics of the chemical reactions and by the kinetics of phase separation

Structures typical for IPNs because of the mechanism of phase separation may appear only within a definite time interval, which is always less than the time of the gelation and of the formation of the final IPN structure. Subsequent (after Δt) cross-linking of both networks or one network occurs in the evolved microregions of phase separation up to when the final conversion degree is reached.

The final structure of IPNs is determined by both thermodynamic and kinetic factors governing chemical reaction and phase separation. Because of such a complicated nature of the simultaneous IPN formation, one can make the following statement about the coexistence of three types of microregions of incomplete phase separation:

(i) Two microregions arise during the phase separation. Their compositions are determined by the time interval Δt and by the temperature of reaction. Each of these microregions is an IPN, which differs in composition from the other one and from the average composition of the system. The presence of these two IPNs with different compositions creates microheterogeneity of the IPN structure

(ii) A nonequilibrium transition region between two phases cross-linked at the final stages of reaction after the phase separation possible under the reaction conditions is attained. The complication in the formation of the IPN structure is in the superposition of two sol–gel transitions in constituent networks during phase separation. At high rates of gelation of one of the networks, the sol–gel transition may precede phase separation of the liquid–liquid type. Therefore, the sequence of phase transitions may vary depending on the system composition and the kinetic parameters of the reaction. Sharp gelation of one of the networks may cause separation according to the nucleation mechanism.

Although there are few experimental data that prove the concept presented above, it may be concluded that the IPN structure is determined by the close interrelation of thermodynamic and kinetic (both chemical and physical kinetics) processes. This means that the thermodynamics of IPN formation and the thermodynamic state of the IPNs are determined to a considerable extent by the chemical kinetics of the cross-linking process.

5.11
Kinetics of IPN Formation in the Presence of Filler

A special problem in the chemistry and physics of IPNs is their reinforcement. This process is tightly connected with the behavior of IPNs at the interface with solid (filler). Physicochemical problems of the filled IPNs were reviewed in [326]. Here, we consider only the filler effect on the kinetics of IPN formation.

As was discussed earlier, the reaction kinetics determines the conditions of phase separation and the segregation degree and fraction of an interphase. From this point of view it is of interest to study the kinetics of IPN formation by introducing fillers. Such investigations were done for simultaneous semi-IPNs made from PU and PBMA in the presence of various amounts of mineral filler (talc, $3MgO \cdot 4SiO_2 \cdot H_2O$) and polymeric filler (cured polyester acrylate) [327, 328].

The kinetic curves are presented in Figs. 75 and 76. The ratios IPN : filler were 80 : 20 and 60 : 40 by mass. As can be seen from Fig. 75, with increasing concentration of initiator autoacceleration of the reaction begins earlier; the conversion degree at the onset of autoacceleration is 0.03–0.05. Introduction of the filler (20 mass %) has no effect on the character of the curves for PBMA at initiator concentrations 5.0×10^{-2} and 2.96×10^{-2} mol·l^{-1} (Fig. 75, curves 1' and 2'). At the same time a decrease in initiator concentration to 0.74×10^{-2} mol·l^{-1} shows some differences on the conversion curve (curve 3'). The onset of autoacceleration is delayed from 160 min with-

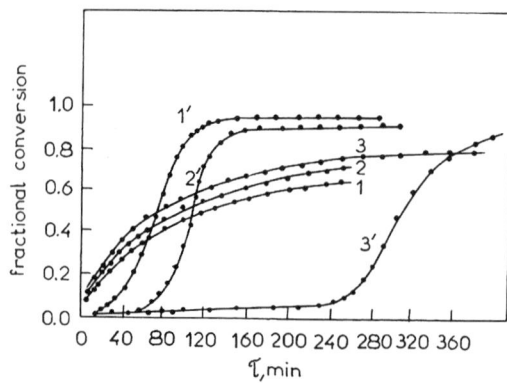

Fig. 75 Conversion curves of PU (1–3) and PBMA (1'–3') formation at 20% filler and initiator concentrations: 5.4×10^{-2} mol·l^{-1} (1), 2.96×10^{-2} mol·l^{-1} (2), and 0.74×10^{-2} mol·l^{-1} (3) [328]

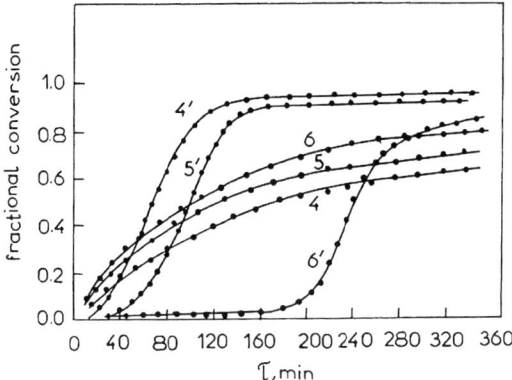

Fig. 76 Conversion curves of PU (4–6) and PBMA (4′–6′) in the presence of 40% of filler at initiator concentrations: 5.4×10^{-2} mol·l^{-1} (4), 2.96×10^{-2} mol·l^{-1} (5), and 0.74×10^{-2} mol·l^{-1} (6) [328]

out the filler to 220 min in its presence, whereas the conversion degree decreases from 0.007 to 0.04. After the onset of autoacceleration, the reaction rate of BMA polymerization decreases in the presence of filler, as is seen from the reduced reaction rate W_{red} (Fig. 77). The dependence of the reduced polymerization rate W_{red} of BMA on the reaction time at various initiator concentrations is also shown: curves 7, 8, and 9 relate to polymerization of pure BMA at initiator concentrations 5.40×10^{-2}, 2.96×10^{-2}, and 0.74×10^{-2} mol·l^{-1}. It is seen that for semi-IPNs with initiator concentrations of 5.40 and 2.96×10^{-2} mol·l^{-1} (curves 1′ and 2′) a decrease of the maximum value without any shift along the time axis is observed, whereas for 0.74×10^{-2} mol·l^{-1} (curve 3′) the maximum becomes smaller and shifts to longer times. As seen from Fig. 75 (curves 1–3), increasing the rate of PBMA formation diminishes the rate of PU formation. By comparing the kinetics of PU formation in the presence of filler with that of unfilled systems [285], it can be seen that the decrease in the reaction rate may be attributed to the dilution of the reaction system by the filler. Increasing the filler content to 40% (Fig. 76) shifts the onset of acceleration to shorter times as compared with unfilled systems. Introducing 40 mass % of filler diminishes the rate of PBMA formation in semi-IPNs for all initiator concentrations (Fig. 77, curves 4′–6′). However, in all cases the influence is more pronounced for the lowest initiator concentration.

The formation of the semi-IPNs proceeds according to two different mechanisms: polyaddition for PU and radical polymerization for BMA. The data show that the introduction of filler influences the kinetics of IPN formation through a single reaction, namely, through radical polymerization.

It is known that in viscous media the role of diffusion processes in polymerization is very important. Diffusion effects may limit both the chain

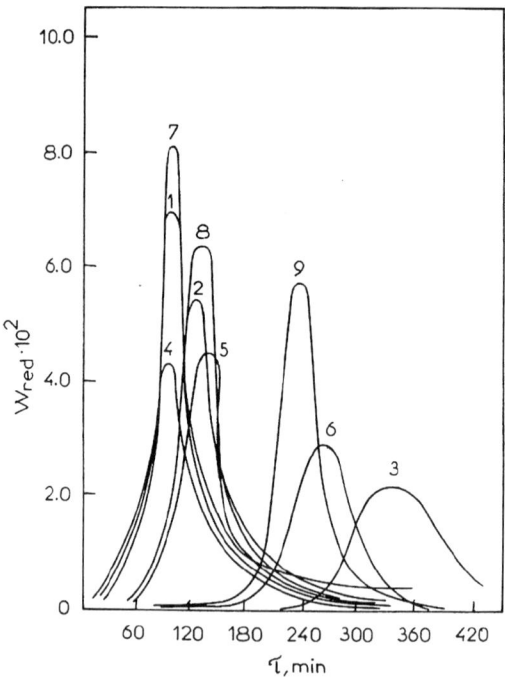

Fig. 77 Dependence of the reduced polymerization rate, W_{red}, of BMA on the reaction time in the presence of filler: 1–3, with 20% filler; 4–6, with 40% filler; 7–9, without filler at various concentrations of initiator $[I] \cdot 10^2$, mol·l^{-1}: 5.40 (1, 4, 7); 2.96 (2, 5, 8); 0.74 (3, 6, 9) [328]

propagation and the initiation. Increasing viscosity leads to a pronounced decrease in the initiator decomposition constant and in the decomposition efficiency. The separation of a radical pair by diffusion is restricted and may enhance the regeneration of the initiator (the so-called cage effect). This effect enables the data for filled IPNs to be explained. Here, the cage effect is revealed by the drop in the reaction rate of PBMA formation at the initial stages of the reaction as compared with the polymerization rate of pure BMA. The effect is less pronounced at higher concentrations of the initiator. This fact was explained in the following way. At the higher initiator concentrations (Fig. 75, curves 1' and 2', and Fig. 76, curves 4' and 5') the formation of PBMA proceeds in a solution of components for PU network formation (the gel point being reached later). At the same time, for this IPN at initiator concentration 0.74×10^{-2} mol·l^{-1}, the rate of PU formation is higher. Also, the PU conversion degree at the onset of autoacceleration of the BMA polymerization reaches the value of 0.7–0.75 and the onset of autoacceleration begins later (180–200 min). These data allow us to conclude that the PBMA formation proceeds in a highly viscous solidlike medium, in which the movements of radicals are impeded, resulting in a sharp decrease in the polymerization rate. Another reason for the appearance of the cage effect may be the local increase in the viscosity of a reaction system at the interface with the solid filler surface, due to adsorption and conformational restrictions of the molecular mobility of growing chains. In principle, the same results were obtained [328] for the same system in the presence of filler—finely dispersed polyester acrylate.

Thus, the conclusion can be drawn that introducing the filler into an IPN hinders phase separation and leads to a permanent structure, in which the segregation and, consequently, the viscoelastic properties differ very much from those of unfilled IPNs. These data agree very well with the effect of the kinetic conditions of IPN formation on the segregation degree discussed above.

The effect of a solid with high surface energy on the IPN formation kinetics was also investigated for another kind of system, the epoxy–allyl IPN produced by simultaneous curing from oligomeric epoxy resin ED-22 and diallyl oligomer products of the condensation of a monoallyl maleinate with ED-20, which has also been studied at the interface with solids of high surface energy [95].

As allyl components for IPN synthesis the following compounds were used: diethylene glycol bis(allyl carbonate) (DEGAC), diallyl phthalate (DAP), diallyl adipate (DAA), and the commercial product of condensation of monoallyl maleinate and epoxy oligomer ED-20. Polymerization of diallyl compounds by formation of IPNs had been performed in the presence of 5 mass % of dibenzoyl peroxide for DEGAC and dicumyl peroxide. As a hardener of epoxy oligomer, complexes of boron trifluoride with an aniline and benzylamine have been used.

The conversion of epoxy and the allylic group concentration in a layer bordering a high-energy surface were estimated by a method of IR–attenuated total reflection (IR-ATR). Compositions with layer thickness of 500 μm were put directly on a prism surface made of KRS-5 glass (incident angle 45°, number of reflections $N = 14$). Under these conditions the depth of radiation penetration is $1-2$ μm. The concentration changes of reactive groups during curing at the interface were compared with those averaged over the bulk. The latter was determined by the usual adsorption IR spectroscopy.

The degree of epoxy and allylic group conversion (or relative fraction of reacted groups) was determined by reducing the changes in the optical density D of the 915 and 1655 cm^{-1} bands for epoxy and allylic groups, respectively, at time t to the optical density D_0 of the same groups at $t = 0$ (5 min after component mixing). Measurements were performed at temperatures of 298, 313, 333, and 353 K.

Using IR spectroscopy, the monotonous decrease of epoxy group concentration during curing of a pure epoxy oligomer ED-20 was found in bulk in the presence of the complex of BF$_3$ with aniline. This was displayed in time by the diminishing 915 cm^{-1} band intensity, corresponding to the asymmetric valent vibrations of the epoxy ring. However, the rate of epoxy group conversion α at 298 K has been found to be very low and after 100 h reaches only

60%. Postcuring of the system during 3 h leads to increasing α up to 87%, and after additional heat treatment at 373 K over 1 h the limiting degree of curing is reached.

At the initial stages of pure epoxy curing in contact with a high-energy surface (glass KRS-5) at 298 K, the total stoppage of epoxy group consumption was observed. However, after additional heat treatment the degree of epoxy group conversion practically did not differ from the values achieved by ED-22 curing in the bulk: conversion α reaches 93% after additional curing at 373 K. Slowing down of the epoxy group consumption at 293 K in the layer bordering a high-energy surface (the layer thickness was 1–2 μm) can be connected with adsorption of the epoxy oligomer and diminishing mobility on the KRS-5 glass surface. It was established that during the simultaneous curing of a mixture of epoxy and diallyl monomers the selective adsorption of an epoxy oligomer at the interface of high surface energy occurs. Due to this fact, full conversion of epoxy groups is not reached. Application of the sequential method of IPN formation or the use during simultaneous IPN formation of an oligomeric diallyl compound diminishes both the distinction in the kinetics of epoxy group conversion and the limiting conversion in the surface layers at the interface of high surface energy and in the bulk.

The influence of two carbon fiber fillers, the basic carbon filler (CF) and one with a surface modified by orthophosphoric acid residuals (PCF), on the kinetics of epoxypolycyanurate (EPCN)–TPU grafted semi-IPN formation and phase structure has been studied [329, 330]. The TPU/EPCN semi-IPNs differed from TPU/PCN ones by the network component synthesized from the reactive blend of EO and DCEBA. The accelerating effect for the reactions occurring is observed by introducing the both carbon fiber fillers into the semi-IPNs studied. The authors [329] connect the acceleration of the chemical processes with the additional chemical reactions between the components and by the catalytic effect of TPU on cyanate cyclotrimerization, as well as by interactions of polymers with the filler surface.

In [308] the filler effect on polymerization kinetics and phase separation in model blends of two linear polymers formed in situ without cross-linking was studied. Blends of PU and PMMA were prepared in the presence of various amounts of fumed silica. It was shown that the filler affects the rates of both reactions. In addition, filler exerts an influence on the phase separation induced by the chemical reaction. Increasing the amount of filler increases the time for the onset of phase separation. The effects observed were explained both by the increase in the viscosity of the reaction system due to introducing filler and by selective adsorption of the reaction components at the interface with filler particles. In all cases, phase separation at the early stages of reaction proceeds in a four-component system (two polymers formed and two initial compounds) and obeys the spinodal mechanism. It was also shown that the final morphology arises far from the end of the reaction and before establishing the equilibrium state.

The general conclusion may be drawn that by formation of IPNs filler plays simultaneously two roles—it affects the reaction rates, which changes the conditions of phase separation, and directly influences the phase separation due to adsorption of components at the interface.

6
Compatibilization in Phase-Separated IPNs

6.1
General Premises

Presently there have been published a great number of works dedicated to the compatibilization of polymer blends. Their main idea consists of reinforcement of the interfacial region between two polymeric phases, namely, of increasing the interaction between components in this region. In most cases this purpose is reached by introduction into the blend of some components (compatibilizers) capable of interacting (physically or chemically) at the interface with both polymers comprising the blend. Compatibilizers are known to improve interphase mixing and interfacial adhesion in incompatible polymer blends [331]. The compatibilizers are supposed to form the intermediate regions between two incompatible phases, being concentrated at their interface. In this case the very term "compatibilization" has more technical character and relates to the processes of modification of interfacial properties of immiscible polymer blends [332]. The introduction of compatibilizers leads to the enhancement of the physical performance of the blend with the formation of a more finely dispersed morphology. Such compatibilization is a way of reinforcing the interface between two immiscible polymers when the added compound (compatibilizer) enhances physical interaction at the interface between the two phases, being concentrated there and interacting simultaneously with both phases via corresponding molecular groups (for example, diblock copolymers). In so doing, the compatibilizer may cause a decrease in the interfacial tension between two homopolymers and results in a reduction in the average size of the dispersed phase [333]. It is supposed that this factor, together with an increase in adhesion between the two phases, results in an improvement in the mechanical properties of the blend. Such compatibilization does not change the thermodynamic stability of the system but changes the morphology [333]. In cited work the interfacial activity of diblock copolymers is described as compatibilizing action, resulting in an increase in the average specific area between the phases. Addition of copolymer can result in complete changes in blend morphology. Thus, the reinforcement of the interface (adhesion at the interface), which is the sign of compatibilization, does not mean an improving thermodynamic interaction between components.

There can be two definitions of the term "compatibilization". The first is the process of reinforcement of the interfacial border in an immiscible polymer blend. The physical essence consists of increasing the interaction between two phases by introduction of substances like diblock polymers, when dissimilar blocks of copolymer will locate at the phase interface. Such compatibilizers do not change the thermodynamic stability of the system and at the same time vary the morphology of the composition, leading to a more fine dispersion of components in the course of blending.

The second definition is based on thermodynamic principles and means increasing the mutual miscibility of two polymeric components [55]. The measure of true thermodynamic compatibility is decreasing the free energy of mixing or thermodynamic interaction parameter between two polymers as a result of introducing compatibilizing substances. To improve the true miscibility of two polymeric components, the added compatibilizer should be distributed throughout the whole volume of the system and in such a way improve the thermodynamic interaction between the blend components and increase the thermodynamic stability of the system. Generally, to realize the miscibility it is necessary that one of the components is miscible with the two other (negative value of χ according to Eq. 1). Thus, it is reasonable to distinguish between two types of compatibilization. In the first case we relate to nonequilibrium compatibilization, whereas the equilibrium case is the process of increasing thermodynamic stability of the polymer–polymer system by introducing a third component [334].

However, in many cases the most important result of introducing compatibilizers is the disappearance of two relaxation maxima on the curves of the dependence of the dynamic mechanical loss tangent on temperature. This disappearance may be the result either of improving the true compatibility or of the morphological changes. It is known that the dynamic mechanical loss tangent is influenced by the morphology of the material, the nature of the interface, and the coupling between the domains of the different phases [335]. The positions of transition and their magnitudes are greatly influenced by the morphology of a material, which depends on the conditions of sample preparation. The slope between the transition regions is influenced by the interfacial regions of varying compositions.

Usually, the presence of two T_g values for polymer blends is taken as an indication of immiscibility. The presence of a single T_g value dependent on the component ratio may signify that either the system is really miscible and consists of one phase, or the domain size in the system is below 15 nm, whereas the system preserves its two-phase state. Thus, the transfer from the system with two glass transition temperatures to the system with a single glass transition temperature may be considered, according to both definitions of compatibilization, as a sign of compatibilization and the only question is what process is under discussion.

The IPNs, as a rule, undergo in the course of curing nonequilibrium microphase separation due to thermodynamic incompatibility of the constituent networks. IPNs present two-phase systems with phases of varying composition. All such systems are characterized by the existence of two relaxation maxima corresponding to the constituent phases. Up to now there are only scattered works where the compatibilization of IPNs is considered. The improvement of compatibility of two networks comprising IPNs may be reached either by grafting one network to the other [148, 336–339] or by introducing compatibilizers of the same type, as were used for blends of linear polymers [339, 340].

Grafting reactions are realized by introduction into the initial reaction mixture of heterofunctional monomers or other compounds able to bind the chains of different networks together. In this case both networks are chemically bound via any intermediate monomer and, because this bonding may be realized only in the course of IPN formation, the compatibilizing agent of such a type should be uniformly distributed in the reaction medium. This means that in the case of grafting one cannot speak about any reinforcement of the interface, because the reaction proceeds before phase separation in the system. Another way of compatibilization of IPNs may be based on the same principles that are developed for polymer blends. In [340] for IPNs based on cross-linked PU and PS, different methods of improving the miscibility were investigated. These included the variation of the cross-links level in both networks, the controlled introduction of internetwork grafting, and the incorporation of various compatibilizers into the PS network.

6.2
Factors with Influence on Determining IPN Compatibility

The compatibility may also be the result of sequential curing, changing the sequence [18, 269, 301] of network formation, and due to specific interaction between the oppositely charged groups or due to formation of hydrogen bonds [341–343, 345, 346], changing the reaction kinetics (the ratio of the rate of formation of constituent networks) [18, 295, 347]. It is well established that the kinetics of formation of constituents of IPNs and their compatibility essentially affect the viscoelastic properties, morphology, and phase continuity of IPNs [202, 348, 349]. It was proved [18, 90, 295, 347, 350, 351] that by varying the kinetics one may achieve the compatibility of two networks.

The state of IPNs based on PBA or PBMA and a diallyl network, obtained by stepwise polymerization (at 333 and 398 K) using two initiators [18] and by simultaneous polymerization with one initiator, was studied using the DMA method. It was discovered that in the first case compatible IPNs were formed, which were characterized by the presence of one very broad maximum of tan δ, typically what is observed for IPNs. It was supposed that in this case gross phase separation is impeded due to the peculiar synthesis mode. For the second case only one narrow maximum was discovered, its position lying between the T_g of two homopolymers, as classically found for random copolymers. The authors supposed that in the second case interfacial grafting proceeded. The reason for this could be the transfer of the growing allylic radical onto previously formed PBA or PBMA. Consequently, it was shown that formation of compatible semi-IPNs based on PBA or PBMA and a diallyl network could be entirely by a free radical mechanism.

PU/polyacrylate IPNs with oppositely charged groups were synthesized and investigated by SAXS and TEM methods [345]. The results show that the attraction between positively charged groups and negatively charged groups improved the miscibility of the two components and their damping capacity. IPNs composed of PU and poly(styrene–acrylic acid) (PSAA) containing mutual oppositely charged groups, i.e., tertiary amine groups in PU and carboxyl groups in PSAA, were prepared with different component ratio PU/PSAA [343]. The DMA of such IPNs shows a single transition peak consistent with SEM observations.

Chemical similarity and H bonding between constituents of IPNs based on the methacrylated diglycidyl ether of bisphenol A (MADGEBA) and DGEBA resulted in good compatibility [346]. The miscibility at the molecular level induced a network interlock effect during IPN formation. This network interlock had a significant influence on the curing behavior of the IPNs and viscoelastic properties; a single damping peak (tan δ) is observed.

Frisch et al. [295] have studied the kinetics of formation of simultaneous semi-IPNs based on poly(dimethylsiloxane) urethane and PMMA using DSC with variation of the parameters determining the kinetics. They discovered a change in the phase morphology with a change of the rates of network formation and their cross-linking density. From the experimental data the conclusion has been drawn that, depending on the kinetic parameters of reaction at the same component ratio, one can obtain both compatible and incompatible networks varying only the rate of network formation.

In [90, 347], from the investigation of the reaction kinetics and the kinetics of phase separation for IPNs based on PS and PU, the role of gelation in the process was established. The authors conclude that if the reaction system separates into two phases before the gel point of PU, the semi-IPNs are turbid, whereas if the phase separation proceeds after the gel point the IPNs formed are transparent.

In [350, 351], the IPNs were obtained using the method of RIM from PU and unsaturated polyesters. It was shown that by varying the component ratio and reaction rates a compatible systems may be obtained. The kinetic study of the formation of simultaneous and sequential acrylic–polyurea IPNs [298] in the course of RIM allowed us to establish that by variation of the component ratio and of the sequence of the reaction it is possible to achieve various degrees of interpenetration up to the formation of transparent IPNs with small domain sizes. The degree of transparency of the IPN materials depended on the relative proportions and domain sizes of the components formed during RIM processing, consistent with the differences observed in reaction kinetics.

6.3
Compatibilization in IPNs Based on PU/PBMA

In some of our works [341, 352] we established the effect of various compounds, which could be potentially considered as compatibilizers, on the compatibilization of IPNs together with studying the kinetics of IPN formation in the presence of such additives and the viscoelastic properties. We supposed that combination of kinetic data and data on the viscoelastic properties of IPNs can give an approach to understanding the mechanism of compatibilization.

For this study we chose well-investigated semi-IPNs based on cross-linked PU and linear PBMA with ratios of components (by mass) of 75 : 25 and 50 : 50. Semi-IPNs were synthesized by the method of simultaneous reactions of polyaddition to form the cross-linked PU and radical polymerization of BMA. Components for PU were POPG MM 2000 and the product of interaction of TMP with 2,4-/2,6-TDI in the ratio 1 : 3.

As potential compatibilizers the following additives were used taken according to various principles:

1. Commercial—2-hydroxyethyl methacrylate (HEMA), containing two functional groups (OH group and double bond) capable of chemical interaction with both components of IPNs.
2. Oligourethane dimethacrylate (OUDM) containing two end double bonds and a urethane block with MM 1500 (also a potential cross-linking agent for BMA).
3. Triblock urethane (TBU) containing a urethane block with MM 1500 and two $CH_3 - CH_2 - O - CH_2 - CH_2$ groups, which do not interact chemically with each component of the IPNs.

OUDM was synthesized from macrodiisocyanate (MDI) based on POPG MM 1000, hexamethylene diisocyanate (HMDI), and HEMA in the ratio 1 : 2 at 40 °C up to full exhaustion of isocyanate groups according to the scheme:

$OCN-R-NCO + 2\ HO-CH_2-CH_2-O-C\ (O)-C(CH_3)=CH_2\ \rightarrow$

  MDI   HEMA

$\rightarrow CH_2=C(CH_3)-C(O)-O-CH_2-CH_2-O-C(O)-NH-R-NH-C(O)-O-CH_2-CH_2-O-C(O)-$

$(CH_3)=CH_2$

            OUDM

where R is $(CH_2)_6-NH-C(O)O-[\ -CH_2-CH(CH_3)-O-]_n O(O)C-NH-(CH_2)_6-$

Scheme 2 Synthesis of OUDM [341]

TBU was obtained from HMDI and the monomethyl ester of ethylene glycol (MEE) at a ratio of 1 : 2 in the presence of tin octoate as a catalyst according to the scheme:

$$OCN–(CH_2)_6–NCO + 2\ HO–CH_2–CH_2–O–C_2H_5 \longrightarrow$$

HMDI   MEE

$$\longrightarrow C_2H_5–O–C_2H_4–O\ (O)\ C–NH–(CH_2)_6–NH–C(O)O–C_2H_4–O–C_2H_5$$

TBU

Scheme 3 Synthesis of TBU [341]

All the potential compatibilizers were introduced into the reaction mixture for IPN formation before the beginning of the reaction. To establish the peculiarities of IPN formation in the presence of compatibilizers, special simultaneous kinetic investigations of the formation of PU and PBMA in IPNs were performed using differential calorimetric methods [285].

Under the same reaction conditions for the same systems the effect of compatibilizers on the time of onset of phase separation was studied using a laser light scattering method. The time of the onset of phase separation was determined from the inflection of the curve of the time dependence of the light scattering intensity at the reaction temperature. Consider some experimental results.

6.3.1
Introduction of Internetwork Grafting

It was shown [336–339] that the introduction of grafting agent leads to a diminishing degree of phase separation and to formation of phases with smaller dimensions of phase domains. We have studied the effect of grafting agent, HEMA, on the kinetics of IPN formation [341, 352]. The onset of phase separation in the forming IPN depends on the reaction kinetics, which, in turn, determines the microphase structure. The kinetic data for some compatibilizers are given in Fig. 78. It is seen that introduction into the reaction system of 5 mass % HEMA (curves 2, 2′) increases the reaction rate for both BMA polymerization and PU formation. This effect is supposed to be connected with chemical interactions between functional groups of HEMA and corresponding groups of the IPN components. One can believe that under these conditions the formation of grafting is very possible. At the same time, the possible grafting does not prevent microphase separation. As follows from Table 18, the time of onset of microphase separation in this case is smaller and separation begins at lower degrees of conversion of both components as compared with IPNs formed without additive. As is seen (Table 18), introduction of 5 mass % HEMA leads at 60 °C to the fast appearance of system turbidity (8 min) at conversion degree for PU, $\alpha_{PU} = 0.25$, and for PBMA

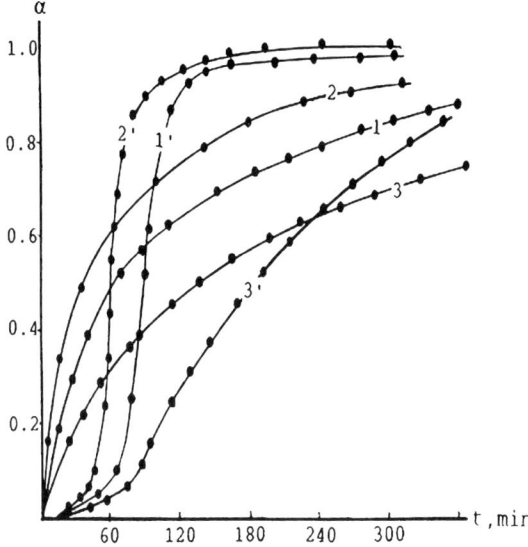

Fig. 78 Kinetic curves for PU (1–3) and PBMA (1′–3′) formation in semi-IPNs of composition 75:25 mass %: initial (1,1′), 5 mass % of MEE (2,2′), 5 mass % of TBU (3,3′) [341]

Table 18 Some features of the compatibilizer action on the formation and properties of IPNs based on PU–PBMA [341]

System PU/PBMA, mass %	Compati- bilizer	Time of the onset of separation, min	Conversion degree at the onset of separation, α		T_g PU, °C	T_g PBMA, °C
			PU	PBMA		
PU	–	–	–	–	– 20	–
PBMA	–	–	–	–	–	75
75 : 25 PU/PBMA	–	56	0.40	0.05	– 5	60
75 : 25 PU/PBMA	5 mass % HEMA	8	0.25	0.01	one T_g = 25	
75 : 25 PU/PBMA	10 mass % OUDM	no separation	–	–	one T_g = 25	
75 : 25 PU/PBMA	5 mass % TBU	60	0.3	0.04	one T_g = 5	
50 : 50 PU/PBMA	–	93	0.38	0.15	0	75
50 : 50 PU/PBMA	5 mass % OUDM	no separation	–	–	0	30

$\alpha_{PBMA} = 0.01$. The experimental curve of the time dependence of light scattering intensity is characterized by a sharp growth of intensity, testifying to the high rate of phase separation. As is seen, introduction of grafting agent into the forming IPN initiates the change in the reaction kinetics of IPN components, which in turn reflects on the onset of microphase separation and viscoelastic properties.

Really, the temperature dependence of the loss maximum $\tan\delta$ for IPN components and semi-IPNs (75 : 25) show the existence of two expressed maxima with glass transition temperatures of $-5\,^\circ C$ for the PU phase and $60\,^\circ C$ for the PBMA phase. This means that the IPN is typically a two-phase system.

For IPNs containing 5 mass % of HEMA, Fig. 79a shows only one broad maximum. The existence of only one maximum may be the result of either compatibilization leading to the formation of a one-phase system or to the formation of a heterogeneous system with broad interphase regions between two phases and with small dimensions of domains of one of the phases. As microphase separation for the systems with introduced HEMA begins much faster than that for the IPN without additive, it is evident that this system

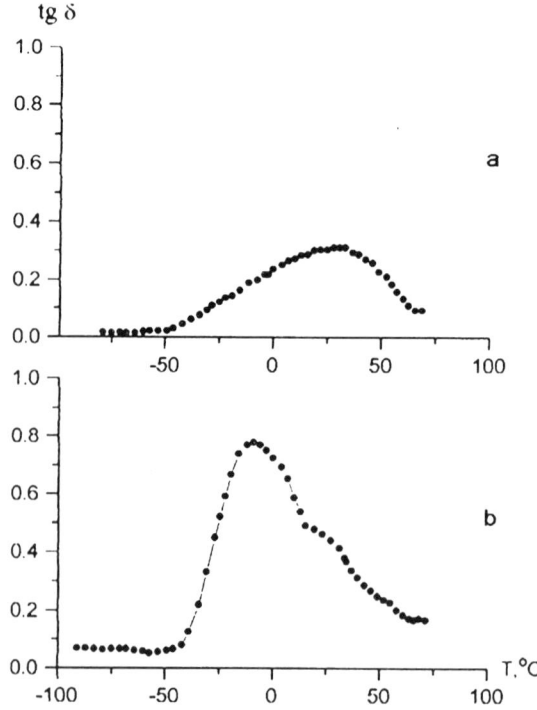

Fig. 79 Temperature dependence of $\tan\delta$ for semi-IPNs (75:25 mass %) with 5 mass % of MEE (**a**) and 5 mass % of TBU (**b**) [341]

is not a one-phase one. On these grounds the conclusion can be drawn that possible chemical grafting, following also from kinetic data, facilitates the formation of the system consisting of small domains and a broad interfacial region.

6.3.2
Introduction of Compatibilizers

It is generally supposed that compatibilizers improve the degree of component mixing and increase the interfacial adhesion between two phases. We have used two compatibilizers: TBU and OUDM [341, 352]. Introduction of 5 mass % of TBU, which does not interact chemically with network components, diminishes both the rates of BMA polymerization and PU formation as compared with the initial semi-IPN (Fig. 78, curves 3, 3′). TBU does not change the time of the microphase separation onset, which, however, begins at lower conversion degrees due to reaction rate retardation. The appearance of turbidity begins after 60 min of reaction. The scattering intensity curve has more stepwise character pointing to a slower rate of microphase separation. The turbidity in this system appears at $\alpha_{PU} = 0.3$ and $\alpha_{BMA} = 0.04$ (Table 18). Introduction of TBU results in the appearance of only one high maximum at $-10\,^{\circ}C$ and a small shoulder in the temperature region $25-40\,^{\circ}C$. A combination of kinetic and mechanical data shows that this IPN is a phase-separated two-phase system where TBU is possibly localized at the phase interfaces, leading to sharp diminishing of the domain sizes. This follows from the chemical constitution of TBU containing blocks identical to IPN components. It is evident that TBU diminishes the degree of phase separation in the system and as a result only one maximum of $\tan\delta$ appears corresponding to a continuous PU phase, which contains discrete inclusions of the second component.

The most interesting results have been obtained when OUDM was used as a compatibilizer. This compound containing two end double bonds may serve as cross-linking agent for BMA. The kinetic curves of polymerization of BMA and of PU formation in the presence of 10 mass % of OUDM show that introduction of OUDM slightly decreases the reaction rates for both components, which may be ascribed to some increase of viscosity of the reaction mixture (OUDM is a viscous liquid soluble in monomer). It is important that in the course of the reaction the system undergoes no phase separation up to the end of reaction. The temperature dependence of the $\tan\delta$ shows only one high maximum with a feebly marked inflection point in the temperature region $45-55\,^{\circ}C$ (Table 18). From these data the conclusion can be drawn that introducing OUDM into the reaction mixture really improves miscibility in such a ternary system. As a result, a ternary compatible IPN is formed. It may be supposed that OUDM physically interacts with both IPN components, like TBU; however, in this case some cross-linking of BMA is also possible. The system is characterized by one relaxation maximum located between the glass transition temperatures of two initially incompatible phases.

Similar results were obtained for IPNs with component ratio 50 : 50 by mass %. It is known that the degree of compatibility of two incompatible polymers depends on the ratio of components. To establish the effect of the component ratio on the compatibilization, the experiments have been done also with 50:50 mass % IPNs. The general regularities for this system are similar to the preceding one. Introduction of 10 mass % of OUDM into 50 : 50 IPN shows only one broad relaxation maximum with a high level of mechanical losses. However, as distinct from 75 : 25 IPN, this maximum is located not between the glass transition temperatures of components, but in the region of the PU component.

6.4
Compatibilization in IPNs Based on PU/PS

Research into PU/PS IPNs was focused on the improvement of compatibility of two networks by use of three different compatibilizers [339, 340]. The compatibilizers used in this study can best be represented as a kind of grafted compatibilizer with excellent anchorage, since both ends were incorporated into the PS network. The three different compatibilizers comprised, first, a very short-chained one consisting of the 1,1,3,3-tetramethylxylene diisocyanate (m-TMXDI) molecule terminated with HEMA (1). This compatibilizer is similar in structure to the hard segment of the PU. The second compatiblizer (2) consisted of a POPG MM 1025 molecule terminated with TMI. The third one (3) was TMXDI and one POPG MM 1025.

All compatibilizers were incorporated at a constant level of 10 mass %. Compatilizer 1 did not have any significant compatibilizing effect. The peak locations of the PU and the PS transitions did not move and the tan δ values of the inner transition region were found to remain low. However, the size and height of the transition peaks suggest that an inversion of the continuous phase took place. This indicates that the 10 mass % compatibilizer went exclusively into the PS network. Consequently, because of the higher quantity of PS in the composition and the increased cross-link density, the PS peak was bigger than the PU peak.

The high molar mass compatibilizer 3 also resulted in inverted tan δ peak sizes. While the PU transition remained in place, a substantial inward shift of the entire PS transition could be noticed. This suggests that compatibilizer 3 caused some degree of component mixing, resulting in less pure PS domains. With respect to the damping ability of the material, no improvement was made, since the intertransition values also remained low.

A surprising result was obtained with compatibilizer 2. The PS T_g was seen to split into two transitions, one at 120 °C and the other at 70 °C. Also, the intertransition regions between the three peaks were high, so that the transition exhibited almost a rectangular shape. Thus, a good damping material with tan δ values > 0.3 at over 135 °C was obtained.

The different performance of the three compatibilizers can be explained by considering that the PU transition stemmed from the soft segments, the POPG chains. Compatibilizer 1 was inefficient since it did not have any affinity for the soft segment and was also very short. The high molar mass compatibilizer 3 was probably dispersed within the PU/PS interphase and the PS domains. Some interaction between the π electrons of the aromatic rings of the diisocyanate and the PS might have decreased the immiscibility with the latter. Compatibilizer 2 probably increased and broadened the interphase region around the PS domains, in this way creating the third transition peak. The PS peak itself was shifted inward to a much lesser degree than for compatibilizer 3. This indicates that, because of the strong immiscibility of the POPG with the PS, only a small fraction of compatibilizer was dispersed in the relatively pure PS domains.

Interesting results were obtained by Hourston et al. [339, 340] who studied the compatibilization of IPNs based on immiscible polymer pairs such as PU and PS. They have shown that introducing various heterofunctional monomers or compatibilizing agents allows regulation of the degree of compatibility from a full one to the system that is characterized by the very broad temperature transition.

Reaction compatibilization by introducing compatibilizing agents was thoroughly investigated for the semi-IPNs based on cross-linked PU and PS in the presence of HEMA and OUDM [310, 311, 353, 354]. Semi-IPNs were obtained by the simultaneous curing of cross-linked PU in the presence of styrene at 333 K. PU was based on the MDI consisting of 2,4/2,6-TDI, POPG MM 1000, and TMP as cross-linking agent. The PU/PS ratio was 70 : 30, 50 : 50, and 30 : 70 by mass.

It was shown that introduction of various amounts of HEMA into the starting system leads to the formation of a semi-IPN which is characterized by a single temperature transition based on DMA and DSC data [310, 353]. The position of this transition depends on the system composition and on the kinetic conditions of the reaction (rates of formation of both components). The kinetic measurements have shown that during reaction in the presence of HEMA no phase separation proceeds, as follows from the light scattering data. In this case the final system has a one-phase structure due to the formation of a thermodynamically miscible ternary system and to the grafting of PU chains onto PS macromolecules via the third component. The reaction compatibilization was studied more thoroughly for semi-IPN PU/PS in the presence of OUDM [311].

The effects of compatibilizing additive OUDM on the kinetics of IPN formation based on cross-linked PU and linear PS and its influence on microphase separation and the viscoelastic and thermophysical properties have been investigated [311, 354]. It was established that 10 and 20 mass % of

OUDM introduced into the initial reaction system prevent microphase separation of the system and lead to the formation of compatible IPNs, as follows from the data on light scattering. The viscoelastic and thermophysical properties of modified IPNs (20 mass % OUDM) are changed in such a way that instead of two relaxation transitions characteristic of phase-separated systems, only one relaxation transition is present. It is the result of a change in the morphology of the system. The position of this relaxation transition depends on the system composition and on the reaction conditions. It was discovered that introduction of OUDM into the semi-IPN changes the kinetics of formation of the systems. These changes affect the time of the onset of phase separation. Introduction of a small amount of OUDM (2.0 and 5.0 mass %) into the reaction system delays the microphase separation. Both the conversion degree of styrene and the degree of the reaction completeness of urethane formation at the onset of phase separation grow compared to the initial system. Increasing the OUDM amount (10.0 and 20.0 mass %) has as a consequence the lack of phase separation. It seems to be possible that increasing the additive amount increases the total number of hydrogen bonds between OUDM and PU, whereas at the same time methacrylate groups partially cross-link PS and partially lead to the formation of branched PS. These factors may improve the compatibility [311].

In [310, 354] the effect of various compatibilizing agents was studied for semi-IPNs based on PU and PS. As potential compatibilizing agents, besides the two already mentioned compounds, oligobutadiene diol rubber and oleic acid were studied. These compounds alter the reaction kinetics, thus affecting the phase structure and viscoelastic properties, but do not convert the modified systems into one phase. The latter results and other data are interesting also from the following point of view. They show the possibility of regulating the viscoelastic properties, changing the segregation degree, positions of two relaxation maxima, and their height. The latter allows us to regulate the damping properties of IPNs.

6.4.1
DMA Investigation

To estimate the compatibilizing action of OUDM on the viscoelastic properties of IPNs, we have selected the semi-IPN based on cross-linked PU and PS at a ratio of components of 70:30 mass % [311]. Figure 80 shows the temperature dependencies of $\tan \delta$ for the initial semi-IPN PU/PS (a) and for semi-IPNs containing 5 (b), 10 (c), and 20 mass % of OUDM (d). Values of T_g are given in Table 19. It is seen from Fig. 80a, that the initial semi-IPN with component ratio 70:30 mass % is characterized by two sharp maxima corresponding to two phases enriched in one of the components. For IPNs with 5 and 10 mass % of OUDM these two maxima are preserved, although the shape of the relaxation curves changes. The approaching, broadening, and dimin-

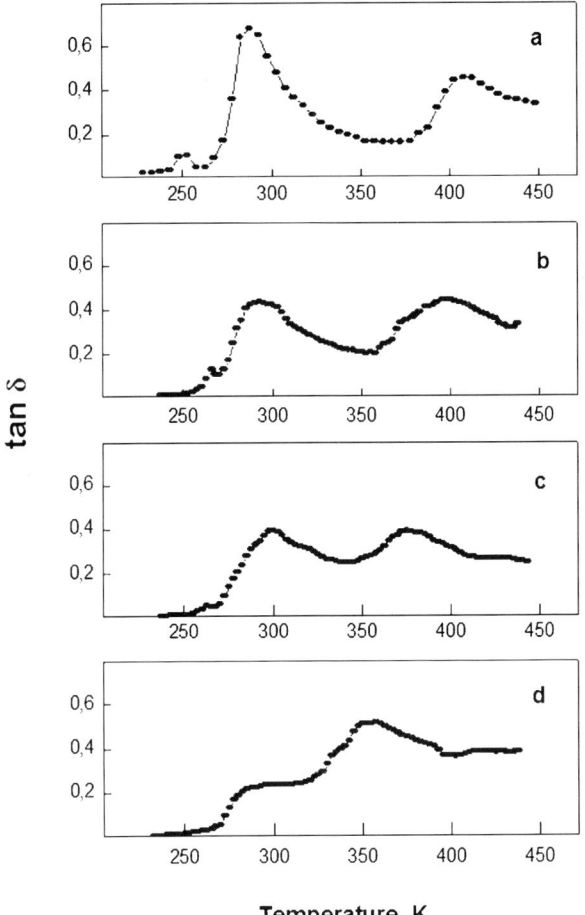

Fig. 80 Temperature dependence of mechanical losses for semi-IPNs PU/PS 70:30 mass % ($[kt] = 0.3 \times 10^{-5}$ mol·l^{-1}, $[I] = 1 \times 10^{-2}$ mol·l^{-1}) with OUDM, mass %: **a** 0, **b** 5, **c** 10, **d** 20 [311]

ishing of the height of the relaxation maxima proceed simultaneously. With increasing amount of OUDM, the glass transition of the PU phase grows from 288 K for IPNs without compatibilizer up to 300 K for the IPN with 10 mass % of OUDM. The T_g of the PS-enriched phase under the same conditions diminishes from 408 to 375 K (Fig. 80a–c; Table 19).

Diminishing of the heights of the relaxation maxima of PU and PS constituents, their approaching and broadening, as well the increasing level of losses between maxima may testify to diminishing amounts of the two phases due to transition of some part of PU and PS into the interfacial region as a result of interaction with OUDM.

Table 19 Parameters of relaxation transition in DSC and DMA for initial and modified semi-IPNs [311]

PU/PS, mass %	OUDM, mass %	DSC T_g, K PU	DSC T_g, K PS	ΔC_p, J·g⁻¹·K⁻¹ PU	ΔC_p, J·g⁻¹·K⁻¹ PS	1−F, %	DMA T_g, K PU	DMA T_g, K PS	tan δ PU	tan δ PS	α_{segr}
100:0	–	251	–	0.68	–	–	283	–	0.88	–	–
100:0	5	265	–	0.65	–	–	288	–	0.94	–	–
100:0	20	255	–	0.70	–	–	286	–	1.06	–	–
0:100	–	–	368	–	0.42	–	–	388	–	3.26	–
0:100	20	–	355	–	0.35	–	–	380	–	1.77	–
70:30	–	258	373	0.55	0.37	17.6	288	408	0.68	0.46	0.31
70:30	2	255	368	0.50	0.25	29.3	–	–	0.44	0.44	–
70:30	5	255	368	0.50	0.20	31.6	293	398	0.39	0.39	0.18
70:30	10	255ᵃ	–	0.60ᵃ	–	–	301	375	0.52ᵃ	–	0.10
70:30	20	255ᵃ	–	0.60ᵃ	–	–	358ᵃ	–	–	–	–
50:50	–	258	370	0.40	0.36	30.9	293	418	0.59	0.65	0.33
50:50	20	275ᵃ	–	0.50ᵃ	–	–	358ᵃ	–	0.89ᵃ	–	–
30:70	–	258	368	0.35	0.35	29.7	303	413	0.43	0.88	0.35
30:70	20	315ᵃ	–	0.30ᵃ	–	–	373ᵃ	–	1.00ᵃ	–	–

ᵃ One relaxation transition is observed

An important characteristic of the two-phase systems is the degree of segregation α_{segr}. Calculation of α_{segr} has shown that an increasing amount of OUDM essentially diminishes this value from 0.31 for IPNs without compatibilizer up to 0.10 for the IPN containing 10 mass %. This means that the introduction of the compatibilizing additive OUDM in the course of chemical reactions of IPN formation prevents microphase separation in the system, i.e., increases the compatibility.

The most remarkable change in the relaxation properties occurs under introduction into the reaction system of 20 mass % of OUDM (Fig. 80d; Table 19). In this case instead of two maxima one observes only one broad maximum at 358 K, whereas the maximum corresponding to the PU phase degenerates and is revealed only as a shoulder (Fig. 80d). Such changes in the viscoelastic behavior may also be the result of morphological transformation [331]. For the formation of semi-IPNs in the presence of 20 mass % of OUDM, the compatibility in the system increases.

It is evident that the compatibilizing effect should depend on the ratio of the components in the IPN. Figure 81 presents the temperature dependencies of the mechanical losses for PU/PS semi-IPNs of various compositions, i.e., 70 : 30 (curve 1), 50 : 50 (curve 2), and 30 : 70 (curve 3), and for the same compositions but formed in the presence of 20 mass % of OUDM (curves 1′, 2′, and 3′, respectively). It follows that the initial IPNs are two-phase systems (curves 1–3). Calculations of the degree of segregation have shown slight increasing of the value α_{segr} with diminishing amount of PU (Table 19).

Introduction of 20 mass % of OUDM in all cases changes the relaxation properties (curves 1′–3′), especially for PU/PS compositions 50 : 50 (curve 2′) and 30 : 70 (curve 3′). Instead of two maxima, these systems reveal only one maximum situated on the temperature scale between temperatures of relaxation transitions of separated phases in the initial IPNs. For these modified IPNs, microphase separation was not observed (Table 19). As is clearly seen from Fig. 81, the height of the maximum and its temperature position depend on the component ratio.

It is clear that all the changes described here testify to the formation of one-phase compatible systems after introduction of a large amount of compatibilizer, although the final conclusion can be drawn only after thermodynamic measurements of the free energy of mixing. It is worth noting that increasing the amount of OUDM changes the properties of IPNs stepwise. However, if the small amounts of additive may be considered as a compatibilizer, the greater amounts (10, 20 mass %) may hardly be described in the same way. In reality the IPNs with such amounts of additive are three-component IPNs, and the changes in their properties are determined by their chemically different structure in relation to the initial two-component IPNs.

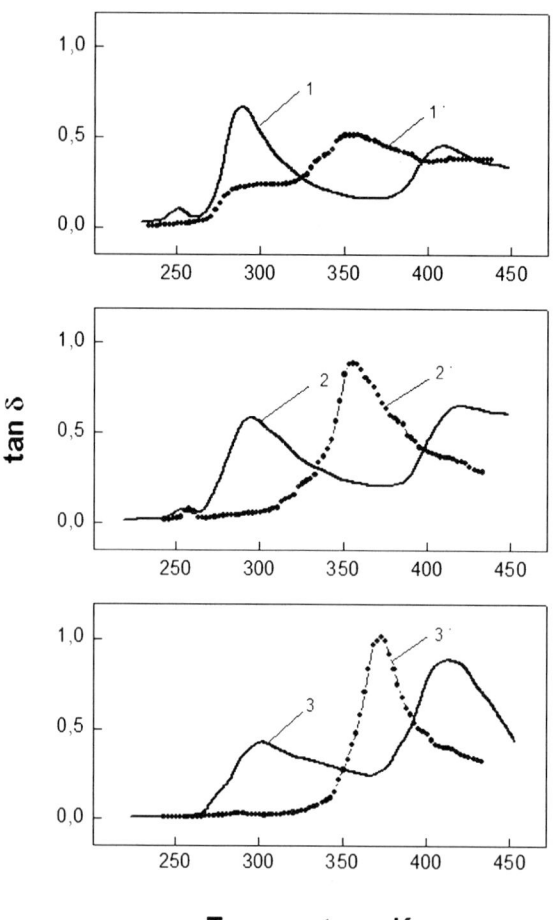

Temperature, K

Fig. 81 Temperature dependence of mechanical losses for initial semi-IPNs (1–3) and semi-IPNs with 20.0 mass% of OUDM (1′–3′) at various component ratios PU/PS, mass %: (1,1′) 70 : 30, (2,2′) 50 : 50, (3,3′) 30 : 70 ([kt] = 0.3 × 10^{-5} mol·l^{-1}, [I] = 1 × 10^{-2} mol·l^{-1}) [311]

6.4.2
DSC Measurements

It is known that the temperature dependence of heat capacities of two-phase polymer systems reveals two transition regions corresponding to the coexisting phases. The glass transition regions are divided by the region of nonlinear change of heat capacity, the latter being ascribed to the smoothly changing composition profile of the interfacial regions (interphase) [99]. The compatibilization of heterogeneous blends usually resulted in the formation of interfacial regions with specific characteristics. The thickness of the inter-phase is estimated as a few nanometers. These questions were thoroughly studied by Eklind [355].

Introduction of compatibilizers into the two-phase system also increases the specific area between the phases and may result in a complete change in blend morphology [333]. Diminishing the dimensions of the phase region is accompanied by an increasing fraction of interfacial region. In such a way, the increasing compatibility in polymer blends by compatibilization may be connected to the formation of an interfacial region between two coexisting phases. The higher the fraction of this region, the higher should be the compatibility.

Studying the thermophysical properties allows us to calculate this fraction from the data on the heat capacity increments at temperature transition [356]. Hourston et al. [98] have used the Freed equation for this purpose. Experimental data for the system are presented in Table 19. It is seen that the values of increments of heat capacity for each phase and temperature position of the transition depends on the ratio of components in the IPNs. DSC data show that the values of increments of heat capacities for initial PS and PU are higher than for the same constituents in IPNs (Table 19), in agreement with the data for polymer blends [100]. Increasing the amount of PS in the IPN leads to growth of the fraction of PS in the interfacial region, whereas a change of the PU fraction has a nonlinear dependence on composition, showing the complicated character of the distribution of IPN components in the interphase. However, for all the systems growth of the content of PS increases the value of $(1 - F)$ (Table 19), which may indicate improvement of the compatibility.

As follows from the experimental data (Table 19), introduction of small amounts of OUDM (2 and 5 mass %) into IPN 70:30 mass % results in an increasing value of $(1 - F)$, i.e., the fraction of the interfacial region increases, testifying to the increasing system compatibility. Simultaneously the fractions of PU and PS in the interfacial region increase 1.3 and 4 times as compared with the initial IPN.

However, by introduction of a higher amount of OUDM (10, 20 mass %) only one jump in the heat capacity occurs (Table 19). It may be supposed that with increasing OUDM amount, a larger fraction of PS transits into the interfacial region and the PS content in the PU-enriched phase becomes less. On increasing the PS content in the IPN, T_g shifts to higher temperatures and the heat capacity increment diminishes (Table 19).

The thermophysical characteristics of the blends containing 20 mass % of OUDM are markedly changed with component ratio (50 : 50 and 30:70 mass %). Correspondingly, calculations $(1 - F)$ become impossible, as the system consists of only one phase. Such a system is characterized by only one maximum of mechanical losses. The appearance of only one relaxation maximum with the increasing amount of compatibilizer may be interpreted as the transition of all the polymers into the "interphase" state, i.e., the border between two phases fully disappears.

Thus, the experimental data on the introduction into the phase-separating IPNs of compatibilizing agent allow the following conclusions to be drawn. Compatibilizer exerts marked effects on the kinetics of formation of both constituents of the IPN, and in such a way changes the conditions of the phase separation during the chemical process. Both the time of the onset of the microphase separation and its rate depend on the amount of compatibilizer and on the ratio of the IPN components. The degree of segregation of the IPN components depends on the reaction kinetics, and its change under compatibilization inevitably affects this value. Successively increasing the amount of compatibilizer leads to the rapprochement of the position of two relaxation maxima appearing in the phase-separated system. Besides, introduction of the compatibilizer, increasing the interaction between the two IPN components, decreases the segregation degree more the higher the amount of compatibilizer. This effect depends also on the ratio of the IPN components.

The compatibilizing effect reveals itself also in the increasing fraction of the interfacial region existing between two phases in the systems where the phase separation is not completed. The fraction of the interfacial region increases with increasing amount of the compatibilizer.

All the IPNs containing 20 mass % of OUDM do not show any sign of phase separation, as follows both from the measurement of the light scattering during reaction and from the single relaxation maximum. Correspondingly, such systems, which may be considered as fully compatible, have no interfacial region and no component segregation. We suppose that the effect of full compatibilization by the introduction of a large amount of compatibilizer is connected not only with strongly improving the interactions at the interface between two phases, but also with the change of the thermodynamic interactions in the ternary system component I–component II–compatibilizer, which may lead to diminishing of the free energy of mixing in the ternary system.

Thus, the investigation of the effects of some additives, which according to their chemical constitution could be considered as potential compatibilizers for IPNs, allows us to establish two different types of compatibilizing action. In one case, the additive introduced into initial reaction system prevents microphase separation of the system. This may mean that the ternary miscible system arises where thermodynamic interactions between components lead to a diminishing total thermodynamic interaction parameter between components down to the negative value typical of thermodynamically stable systems. In this case, the viscoelastic properties are changed in such a way that instead of two loss maxima characteristic of phase-separated system, only one maximum is present, which is possibly a result of the formation of a one-phase ternary system.

In the other case, the additives exert some effect on the reaction kinetics and because of it accelerate microphase separation in the system, which proceeds at lower conversion degrees as compared with pure IPNs. Under such conditions, thermodynamic compatibility of the IPN components in the presence of additives decreases. The acceleration of the reaction in the presence of additives only shortens the time for the onset of phase separation. However, in this case there proceeds also a transition from the systems with two relaxation maxima (two-phase system) to the system with one, broad or narrow, maximum. This may be the result of concentration of the additive in the interfacial region between two phases. It is evident that the morphology of such a system also undergoes changes and that instead of discrete phases, a diffuse structure or structure of the matrix-inclusion type is formed. In such structures the sizes of the phase domains (inclusions) of the component present in smaller amounts are rather small to be detected by methods of DMS. In fact, detection of a single T_g only signifies that the size of the blend domains is very small [331, 332].

The effect of additives on the kinetics of formation of the IPN components and on the time of onset of phase separation may serve as a test of the compatibilization mechanism. The additive is introduced into the homogeneous reaction mixture and so is distributed uniformly throughout the whole volume of it. If the additive acts as a third component improving miscibility, the onset of the phase separation should be shifted on the timescale to higher times or should be fully absent. On the contrary, if the additive has no effect on miscibility or has a negative effect (increasing the thermodynamic interaction parameter of the ternary system), one should suppose that this additive is concentrated at the interface between two phases, changing the morphology of the system that is formed in the course of IPN curing.

This effect is of importance, as it was shown that the reaction rates determine the time of the onset of phase separation, the degree of phase separation, and the fraction of an interphase. Correspondingly, the viscoelastic properties of compatibilized IPNs depend not only on the presence of the compatibilizing agent, but also on its effect on the reaction kinetics.

From the experimental data presented above it follows that, as a rule, the introduction of compatibilizers increases the fraction of an interphase, which may be interpreted as apparently increasing compatibility. However, the physical essence of this effect is not connected to real improving compatibility, i.e., to increasing thermodynamic stability of the system. Generally, the formation of an interphase is the result of uncompleted phase separation by which some part of the system stays in an unseparated (mixed) state partially corresponding to the state of the system before the onset of phase separation (frozen compatibility). In such a way, the interphase is the region of the thermodynamic instability and increases the thermodynamic instability of the system. It is evident that the introduction of compatibilizing agents prevents phase separation and increases that part of the system which is not phase-separated. Compatibilizer transfers the system to the less stable state by increasing the fraction of the system which is not separated.

References

1. Sperling LH (1981) Interpenetrating polymer networks and related materials. Plenum, New York
2. Millar JR (1960) J Chem Soc 3:1311
3. Sperling LH (1994) IPN around the world, II: recent advances. In: Klempner D, Frisch K (eds) Advances in interpenetrating polymer networks, vol 4. Technomic, Lancaster, PA, p 1
4. Sperling LH, Mischra V (1996) Polym Adv Technol 76:197
5. Frisch HL, Wasserman E (1961) J Am Chem Soc 83:3789
6. Frisch HL, Klempner D (1970) Adv Macromol Chem 2:149
7. Frisch HL, Klempner D, Frisch KC (1969) J Polym Sci B 8:775
8. Klempner D, Frisch HL, Frisch KS (1970) J Polym Sci A 7:921
9. Klempner D, Frisch KC (1974) Advances in urethane science and technology. Technomic, Lancaster, PA, p 14
10. Sperling LH, Friedman DW (1969) J Polym Sci A2 7:427
11. Lipatov YS, Lipatova TE, Kosyanchuk LF (1989) Adv Polym Sci 88:50
12. Vilgis TA, Frisch HL (1989) Polym Bull 21:655
13. Lipatov YS, Nizel'sky YN (1993) New J Chem 17:1144
14. Frisch HL (1993) New J Chem 17:697
15. Sperling LH (1986) Recent developments in interpenetrating polymer networks and related materials. In: Paul DR, Sperling LH (eds) Multicomponent polymer materials. Adv Chem Ser 211, American Chemical Society, Washington, DC, p 21
16. Sperling LH (1994) In: Klempner D, Sperling LH, Utracki LA (eds) Interpenetrating polymer networks. Adv Chem Ser 239, American Chemical Society, Washington, DC, p 3
17. Lipatova TE, Shilov VV, Bazilevskaya NP, Lipatov YS (1977) Br Polym J 9:159
18. Derrough SN, Rout C, Widmaier JM (1993) J Appl Polym Sci 48:1183
19. Bauer BJ, Briber RM, Dickens B (1994) In: Klempner D, Sperling LH, Utracki LA (eds) Interpenetrating polymer networks. Adv Chem Ser 239, American Chemical Society, Washington, DC, p 179
20. Gul V (1978) Structure and strength of polymers. Khimiya, Moscow (in Russian)
21. Flory P (1944) Chem Rev 35:51
22. Lipatov YS, Kercha YY, Sergeeva LM (1970) Structure and properties of polyurethanes. Naukova Dumka, Kiev (in Russian)
23. Lipatova TE (1974) Catalytic polymerization and formation of polymer networks. Naukova Dumka, Kiev (in Russian)
24. Irzhak VI, Rosenberg BA, Enikolopyan NS (1979) Crosslinked polymers. Nauka, Moscow (in Russian)
25. Irzhak VI, Korolev GV, Solov'ev ME (1997) Usp Khim 66:179
26. Korolev GV, Mogilevich MM, Golikov IV (1995) Network polyacrylates: microheterogeneous structures, physical networks and deformation and ultimate properties. Khimiya, Moscow
27. Dušek K (1965) J Polym Sci B 3:209
28. Dušek K (1967) J Polym Sci C 16:1289
29. Scidl J, Malinsky J, Dušek K, Heltz W (1967) Adv Polym Sci 5:113
30. Dušek K (1970) Polymer Prepr 11:536
31. Dušek K, Prins W (1969) Adv Polym Sci 6:1
32. Bobalek E, Moore E, Levy S, Lee C (1964) J Appl Polym Sci 8:625
33. Lipatova TE (1975) Pure Appl Chem 43:27

34. Lipatova TE, Ivaschenko VK (1969) Vysokomol Soedin 11:2217 (in Russian)
35. Nesterov AE, Lipatova TE, Ivaschenko VK, Lipatov YS (1970) Vysokomol Soedin B 12:150
36. Lipatova TE, Zubko SA (1970) Vysokomol Soedin A 12:1555
37. Berlin AA, Kefeli TY, Korolev GV (1967) Polyesteracrylates. Khimiya, Moscow
38. Mark H, Tobolsky A (1950) Physical chemistry of high polymeric systems. Interscience, New York
39. Graessley WW (1982) Adv Polym Sci 47:67
40. Glaessley WW (1974) Adv Polym Sci 16:1
41. Mark JE, Eisenberg A, Glaessley WW, Mandelkern L (1984) Physical properties of polymers. ACS, Washington, DC
42. Porter RS, Jonson JF (1966) Chem Rev 66:1
43. Chee KK (1987) J Appl Polym Sci 33:1067
44. Hoffman M (1967) Rheol Acta 6:377
45. Tonelli AE (1970) J Polym Sci Polym Chem Ed 8:656
46. Frank-Kamenetsky MD, Vologodsky AV (1981) Usp fiziki 139:641
47. De Gennes PG (1979) Scaling concept in polymer physics. Cornell University Press, Ithaca
48. Kluppel M (1991) In: Physics of polymer networks. The 29th Europhysics Conference. Eur Phys Soc, Alexisbad, p 85
49. Lipatov YS (1985) Pure Appl Chem 57:1691
50. Park IH, Lee JH, Kim SC (1983) Polym Bull 10:126
51. Lipatov YS (1990) J Macromol Sci Rev Chem Phys C 30:209
52. Nesterov AE, Lipatov YS (1997) Thermodynamics of polymer blends. ChemTec Publ, Toronto
53. Lipatov YS, Grigoryeva OP, Kovernik G, Shilov VV, Sergeeva LM (1985) Makromol Chem 186:1401
54. Lipatov YS, Nesterov AE, Kuzmina GP, Ignatova TD (1993) Polym Network Blend 3:89
55. Olabisi O, Robeson L, Shaw M (1979) Polymer–polymer miscibility. Academic, New York
56. Sophiea D, Klempner D, Sinjarevic V, Suthar B, Frisch KC (1994) In: Klempner D, Sperling LH, Utracki LA (eds) Interpenetrating polymer networks. Adv Chem Ser 239. American Chemical Society, Washington, DC, p 40
57. An JH, Sperling LH (1988) In: Dickie RA, Labana SS, Bauer RS (eds) Cross-linked polymers: chemistry, properties and application. ACS Symp Ser 367. American Chemical Society, Washington, DC
58. Sperling LH, Heck CS, An JH (1989) In: Utracki LA, Weiss BA (eds) Multiphase polymers: blends and ionomers. ACS Symp Ser 395. American Chemical Society, Washington, DC
59. Mishra V, Du Prez FE, Goethals E, Sperling LH (1995) J Appl Polym Sci 58:331
60. Baidak AA, Liegeois JM, Sperling LH (1997) J Appl Polym Sci Polym Phys 35:1973
61. Du Prez FE, Tan P, Goethals E (1996) Polym Adv Technol 7:257
62. Bauer BJ, Briber RM, Han CC (1989) Macromolecules 22:940
63. Li BY, Bi XP, Zhang DH, Wang FS (1989) Forced compatibility and mutual entaglements in poly(vinyl acetate)/poly(methyl acrylate) IPNs. In: Klempner D, Frisch K (eds) Advances in interpenetrating polymer networks, vol 1. Technomic, Lancaster, PA, p 203
64. Lipatov YS, Nesterov AE, Sergeeva LM (1974) Dokl Akad Nauk SSSR 220:637
65. Lipatov YS, Sergeeva LM, Karabanova LV, Nesterov AE, Ignatova TD (1976) Vysokomol Soedin A 18:1025 (in Russian)

66. Nesterov AE, Lipatov YS (1976) Inverse gas chromatography in polymer thermodynamics. Naukova Dumka, Kiev (in Russian)
67. Lipatov YS, Karabanova LV, Gorbach LA, Lutsyk ED, Sergeeva LM (1992) Polym Int 28:99
68. Tager A (1977) Vysokomol Soedin A 19:1654 (in Russian)
69. Lipatov YS, Karabanova LV, Khramova TS, Sergeeva LM (1978) Vysokomol Soedin A 20:46
70. Karabanova LV, Mikhalovsky SV, Sergeeva LM, Meikle ST, Helias M, Lloyt W (2004) Polym Eng Sci 44:940
71. Cahn J, Hilliard J (1958) J Chem Phys 28:258
72. Donatelli AA, Sperling L (1976) Macromolecules 9:672
73. Yeo JK, Sperling LH, Thomas DA (1982) Polym Eng Sci 22:190
74. Donatelli AA, Sperling LH, Thomas DA (1977) J Appl Polym Sci 21:1189
75. Yeo JK, Sperling LH, Thomas DA (1983) Polymer 24:307
76. Williams JA, Borrajo J, Adabbo HE, Rojas AJ (1984) Adv Chem Ser 1285:195
77. Vaquez A, Rojas A, Adabbo HE, Borrajo J, William PJ (1987) Polymer 28:1156
78. Rosenberg B (1991) Macromol Chem Macromol Symp 41:65
79. Manevich E, Sigalov GM, Rosenberg B (2001) In: Rosenberg B, Sigalov GM, Taylor L (eds) Heterogeneous network polymers. Francis Books, London
80. Rosenberg B (2001) Russ I Mendeleev D Chem Soc 45:23
81. Rosenberg B, Sigalov G (1996) Polym Adv Technol 7:356
82. Lipatov YS, Shilov VV (1984) Usp Khim 53:1197
83. Lipatov YS, Grigoryeva OP, Shilov VV, Sergeeva LM (1984) Dokl Akad Nauk SSSR 278:409
84. Lipatov YS, Rosovitsky VF, Maslak YV, Zakharenko SA (1984) Dokl Akad Nauk SSSR 278:1408
85. Lipatov YS, Grigoryeva OP, Sergeeva LM, Shilov VV (1986) Vysokomol Soedin A 28:385
86. Lipatov YS (1983) Mech Compos Mater 5:771
87. Lipatov YS, Shilov VV, Gomza YP, Kovernik GP (1984) Makromol Chem 185:947
88. Lipatov YS (1989) Microphase separation in interpenetrating polymer networks and their viscoelasticity. In: Klempner D, Frisch K (eds) Advances in interpenetrating polymer networks, vol 1. Technomic, Lancaster, PA, p 261
89. Park JW, Kim SS (1996) Polym Adv Technol 7:209
90. He X, Widmaier JM, Meyer G (1993) Polym Int 32:295
91. Shilov VV, Gomza YP, Kovernik GP (1983) Dokl Akad Nauk SSSR 271:913
92. Bauer BJ, Briber RM (1994) The effect of crosslink density on phase separation in interpenetrating polymer networks. In: Klempner D, Frisch K (eds) Advances in interpenetrating polymer networks, vol 4. Technomic, Lancaster, PA, p 45
93. Binder K, Frisch HL (1984) J Chem Phys 81:2126
94. Lipatov YS, Khramova TS, Sergeeva LM, Karabanova LV (1977) J Polym Sci Polym Chem 15:427
95. Mikhal'chuk VM, Lipatov YS, Stroganoff VF (1996) Polym Network Blend 6:133
96. Fox TG (1956) Bull Am Phys Soc 1:123
97. Hsieh KH, Liao DC, Chen CY, Chiu WY (1996) Polym Adv Technol 7:265
98. Hourston DJ, Song M, Hammiche A, Pollock HM, Reading M (1997) Polymer 38:1
99. Spaans RD, Muhammad M, Williams MC (1999) J Polym Sci Polym Phys 37:267
100. Pang YX, Jia DM, Hu HJ, Hourston DJ, Song M (2000) Polymer 41:357
101. Fried JR (1978) PhD thesis, University of Massachusetts

102. Song M, Zhang J, Li Y, Zhang D, Tang X (1989) Kinetic effects in the formation of castor oil polyurethane/epoxide–episulfide resin interpenetrating polymer networks. In: Klempner D, Frisch K (eds) Advances in interpenetrating polymer networks, vol 1. Technomic, Lancaster, PA, p 21

103. Lipatov YS (1994) In: Klempner D, Sperling LH, Utracki LA (eds) Interpenetrating polymer networks. Adv Chem Ser 239. American Chemical Society, Washington, DC, p 125

104. Helfand E (1982) In: Polymer compatibility and incompatibility. Harwood Academic, London, p 133

105. Flory P, Rehner J (1943) J Chem Phys 11:512

106. Treloar LRG (1958) The physics of rubber elasticity, 2nd ed. Oxford University Press, London

107. Shibayama K, Suzuki Y (1967) Rubber Chem Technol 40:476

108. Klempner D, Frisch HL, Frisch KC (1971) J Elastoplastics 3:2

109. Klempner D, Frisch HL (1970) J Polym Sci A2 8:921

110. Frisch HL, Frisch KC (1969) J Polym Sci Polym Lett 7:775

111. Erman B, Mark JE (1997) Structure and properties of rubberlike networks. Oxford University Press, Oxford

112. Lipatov YS, Sergeeva LM (1979) Interpenetrating polymer networks. Naukova Dumka, Kiev (in Russian)

113. Tadao K, Shiro N, Hirochi A (1990) Sequential interpenetrating polymer networks: synthesis, structure, and properties. In: Klempner D, Frisch K (eds) Advances in interpenetrating polymer networks, vol 2. Technomic, Lancaster, PA, p 47

114. Thiele J, Cohen RE (1979) Polym Eng Sci 19:284

115. Siegfried DC, Thomas DA, Sperling LH (1979) Macromolecules 12:586

116. Liu W, Han X, Liu J, Zhou H (1994) In: Klempner D, Sperling L, Utracki L (eds) Interpenetrating polymer networks. ACS, Washington, DC, p 571

117. Lipatov YS, Sergeeva LM, Mozzhukhina LV, Apukhtina NP (1974) Vysokomol Soedin A 16:2290

118. Lipatov YS, Alekseeva TT (1996) Polym Sci A 38:940

119. Lipatov YS, Alekseeva TT (1996) Polym Adv Technol 7:234

120. Lipatov YS, Feinerman AE (1984) J Colloid Interface Sci 98:126

121. Lipatov YS, Vilensky VA (1975) Vysokomol Soedin A 17:2069

122. Fujita HF, Kishimoto A (1960) Trans Faraday Soc 56:424

123. Lipatov YS, Karabanova LV, Sergeeva LM, Gorichko EY (1982) Vysokomol Soedin A 24:110

124. Lipatov YS, Karabanova LV, Sergeeva LM (1984) Vysokomol Soedin A 26:2265

125. Hill TL (1992) Statistical mechanics. McGraw-Hill, London

126. Feinerman AE, Lipatov YS (1995) Polym Network Blend 5:123

127. Feinerman AE, Lipatov YS (2000) J Polym Sci 38:899

128. Kumar H, Kumaraswamy GN, Ravikumar HB, Ranganathaiah C (2005) Polym Int 54:1401

129. Bonart R, Mueller E (1974) J Macromol Sci B10:177

130. Lipatov YS, Shilov VV, Gomza YP, Kruglyak NE (1982) X-ray methods of studying polymeric systems. Naukova Dumka, Kiev (in Russian)

131. Porod G (1952) Koll Z 125:51

132. Alig I, Junker M, Schulz M, Frisch HL (1996) Polym Sci A 38:43

133. Leibler L (1980) Macromolecules 13:1602

134. Shilov VV, Lipatov YS (1986) In: Physical chemistry of multicomponent polymeric systems, vol 2. Naukova Dumka, Kiev, p 25

135. Lipatova TE, Kuzmenko LS, Shilov VV (1978) Vysokomol Soedin A 20:2013
136. Lipatov YS, Shilov VV, Bogdanovich VV (1987) J Polym Sci Polym Phys 25:43
137. Lipatov YS, Shilov VV, Gomza YP (1983) Angew Makromol Chem 119:173
138. Shilov VV, Lipatov YS, Karabanova LV, Sergeeva LM (1979) J Polym Sci Chem Ed 17:3083
139. Lipatov YS, Nesterov AE, Sergeeva LM (1975) Dokl Akad Nauk SSSR 220:637
140. Lipatov YS, Sergeeva LM, Karabanova LV (1976) Vysokomol Soedin A 18:1035
141. Lipatov YS, Babich VF, Karabanova LV (1976) Rep Acad Sci Ukrainian SSR B1:39
142. Shilov VV, Karabanova LV, David L, Boiteax G, Seytre G, Gomza YP, Sergeeva LM (2005) Polymer J 27:255
143. Lipatova TE, Shapoval GS, Shevchuk ES, Basilevskaya NP (1971) J Macromol Sci Chem A 5:345
144. Bershtein VA, Yakasher PW, Karabanova LV, Sergeeva LM, Pissis P (1999) J Polym Sci Polym Phys 37:429
145. Lipatov YS, Sergeeva LM (1986) Usp Khim 55:2086
146. Lipatov YS, Shilov VV, Bogdanovich VA, Karabanova LV, Sergeeva LM (1982) Rep Acad Sci Ukrainian SSR B4:32
147. Lipatov YS, Alekseeva TT, Shilov VV (1991) Polym Network Blend 1:129
148. Song M, Hourston DG, Schater FU (2001) J Appl Polym Sci 79:1958
149. Sperling LH (1984) Polym Eng Sci 24:2
150. McGarey B (1989) A small-angle neutron scattering study of polydimethylsiloxane/polydeuterostyrene sequential IPNs. In: Klempner D, Frisch K (eds) Advances in interpenetrating polymer networks, vol 1. Technomic, Lancaster, PA, p 69
151. Richards RW (1990) Macromol Chem Macromol Symp 40:209
152. Schmatz W, Springer T, Shelten J, Ibel K (1974) J Appl Polym Sci 7:96
153. Kostorz G (1979) Treatise on material science and technology. Academics, New York
154. Richards RW (1985) Adv Polym Sci 71:1
155. Debye P, Bueche AM (1949) J Appl Phys 20:518
156. Hosemann R (1950) Kolloid Z 117:13
157. Debuy P, Anderson HR, Brumberger H (1957) J Appl Phys 28:679
158. Sperling LH, Klein A, Fernandez AM, Linne MA (1983) Polymer Prepr 24:386
159. Fernandez AM, Wignall GD, Sperling LH (1986) In: Paul DR, Sperling LH (eds) Multicomponent polymer materials. Adv Chem Ser 211. American Chemical Society, Washington, DC, p 153
160. Lebedev EV, Lipatov YS, Bezruk LI (1975) New methods of polymer investigations. Naukova Dumka, Kiev
161. Kim SC, Klempner D, Frisch KC, Radigan W, Frisch HL (1976) Macromolecules 9:258
162. Kim SC, Klempner D, Frisch KC (1975) J Polym Sci 15:340
163. Touhsaent RE, Thomas DA, Spering LH (1976) Adv Chem 154:206
164. Sperling LH (1977) Macromol Rev 12:141
165. Touhsaent RE, Thomas DA, Sperling LH (1974) J Polym Sci C 46:175
166. Sperling LH (1977) Interpenetrating polymer networks and related materials, suppl 1. In: Encyclopedia of polymer science and technology. Wiley, New York, p 288
167. Jordhamo GM, Manson JA, Sperling LH (1986) Polym Eng Sci 26:517
168. Sperling LH, Widmaier JM (1983) Polym Eng Sci 23:694
169. Widmaier JM, Sperling LH (1982) J Appl Polym Sci 27:3513
170. Widmaier JM, Sperling LH (1982) Macromolecules 15:625
171. Widmaier JM, Yeo JK, Sperling LH (1982) Colloid Polym Sci 260:678

172. Jia D, Chen L, Wu B, Wang M (1989) Interpenetrating polymer networks based on polybutadiene-based polyurethane. In: Klempner D, Frisch K (eds) Advances in interpenetrating polymer networks, vol 1. Technomic, Lancaster, PA, p 303

173. Lee BK, Kim SC (1995) Polym Adv Technol 14:402

174. Lipatov YS, Rosovitsky VF, Alekseeva TT, Babkina NV (1988) Dokl Akad Nauk UkrSSR B5:52

175. Fesko D, Tschoegl NW (1971) J Polym Sci C 35:51

176. Takayanagi M, Uemura S, Minami S (1964) J Polym Sci C 5:113

177. Kraus G (1971) In: Multicomponent polymer materials. Adv Chem Ser 99. American Chemical Society, Washington, DC, p 189

178. Frisch KC, Klempner D, Frisch HL, Ghiradella H (1974) In: Sperling LH (eds) Recent advances in polymer blends, grafts and block copolymers. Plenum, New York, p 395

179. Matsuo M, Kwei TK, Klempner D, Frisch HL (1970) Polym Eng Sci 10:327

180. Lipatov YS (1994) Polymer reinforcement. ChemTec, Toronto

181. Kim SC, Klempner D, Frisch KC, Frisch HL (1977) Macromolecules 10:1187

182. Budiansky B (1965) J Mech Phys Solids 13:223

183. Davies WE (1971) J Phys D 4:318

184. Rosovitsky VF, Ilavsky M, Hrouz J, Dusek K, Lipatov YS (1979) J Appl Polym Sci 24:1007

185. Qin J, Li F, Wu Z, Qian B (1990) Interpenetrating polymer networks in acrylic blends and their fibers. In: Klempner D, Frisch K (eds) Advances in interpenetrating polymer networks, vol 2. Technomic, Lancaster, PA, p 205

186. Sperling LH, George HF, Huelck V (1969) J Appl Polym Sci A2 7:425

187. Huelck V, Thomas DA, Sperling LH (1972) Macromolecules 5:340

188. Huelk V, Thomas DA, Sperling LH (1971) Am Chem Soc Prepr 12:665

189. Donatelli AA, Sperling LH, Thomas DA (1972) Macromolecules 9:676

190. Yenwo GM, Sperling LH, Pulido J, Manson JA, Conde A (1977) Polym Eng Sci 17:17

191. Karabanova LV, Boiteax G, Gain U, Seytre G, Sergeeva LM, Lutsuk ED (2004) Polym Int 53:2051

192. Lipatov YS, Babich VF, Karabanova LV, Korghuk NI, Sergeeva LM (1976) Rep Acad Sci Ukrainian SSR B1:39

193. Touhsaent RE, Thomas DA, Sperling LH (1976) Adv Chem 154:206

194. Tsunoda S, Suzuki Y (1990) Interpenetrating polymer networks from epoxy resin and polyacrylate: synthesis, characteristics, and application to electrical/electronic fields. In: Klempner D, Frisch K (eds) Advances in interpenetrating polymer networks, vol 2. Technomic, Lancaster, PA, p 177

195. Lipatov YS, Sergeeva LM (1986) Usp Khim 55:2086

196. Macknight WI, Eatness TR (1981) J Polym Sci Rev 6:41

197. Siegfried DL, Thomas DA, Sperling LH (1981) Polym Eng Sci 21:39

198. Siegfried DL, Thomas DA, Sperling LH (1980) Polymer Prepr 21:186

199. Allen G, Bowden MJ, Blundell DJ (1973) Polymer 14:604

200. Allen G, Bowden MJ, Lewis G (1974) Polymer 15:13

201. Allen G, Bowden MJ, Lewis G (1974) Polymer 15:19

202. Lipatov YS, Alekseeva TT, Rosovitsky VF, Babkina NV (1992) Polymer 33:610

203. Manoj NR, Raut RD, Sivaraman P, Ratna D, Chakraborty BL (2005) J Appl Polym Sci 96:1487

204. Dadbin S, Frounchi MH (2003) J Appl Polym Sci 89:1583

205. Vancaeyzeele C, Fichet O, Boileau S, Tayssie D (2005) Polymer 46:6888

206. Dadbin S, Chaplin RP (2001) J Appl Polym Sci 81:3361

207. Mathew A, Packirisamy S, Thomas S (2000) J Appl Polym Sci 78:2327

208. Samui AB, Dalvi VG, Patri M, Chakraborti BC, Deb PC (2003) J Appl Polym Sci 91:354
209. Ali SA, Hourston DI (1994) Thermoplastic interpenetrating polymer networks. In: Klempner D, Frisch K (eds) Advances in interpenetrating polymer networks, vol 4. Technomic, Lancaster, PA, p 17
210. Fainleib A, Grigoryeva O, Pissis P (2005) J Balkan Tribol Assoc 11:303
211. Grigoryeva O, Fainleib A, Sergeeva LM (2005) Thermoplastic polyurethane elastomers in interpenetrating polymer networks. In: Fakirov S (ed) Handbook of condensation thermoplastic elastomers. Wiley-VCH, Weinheim, p 325
212. Vatalis A, Delides C, Grigoryeva O, Sergeeva L, Brovko A, Zimich O, Shtompel V, Georgoussis G, Pissis P (2000) Polym Eng Sci 40:2072
213. Vatalis A, Delides C, Georgoussis G, Kyritsis A, Grigoryeva O, Sergeeva L, Brovko A, Zimich O, Shtompel V, Neagu E, Pissis P (2001) Thermochim Acta 371:87
214. Won HJ, Yong KK, Ick HK (1991) Macromolecules 24:4708
215. Hermant I, Damyanidu M, Meyer GC (1983) Polymer 24:1419
216. Rizos AK, Fytas G, Ma RJ, Wang CH, Abetz V, Meyer GC (1993) Macromolecules 26:1869
217. Pandit SB, Nadkarni VM (1994) Macromolecules 27:4583
218. Sergeeva L, Grigoryeva O, Zimich O, Privalko E, Shtompel V, Privalko V, Pissis P, Kyritsis A (1997) J Adhesion 64:161
219. Sergeeva L, Grigoryeva O, Brovko A, Zimich O, Nedashkovskaya N, Slinchenko E, Shtompel V (1997) J Prikladn Khim 70:2038 (in Russian)
220. Kyritsis A, Pissis P, Grigoryeva O, Sergeeva L, Brovko A, Zimich O, Privalko E, Shtompel V, Privalko V (1999) J Appl Polym Sci 73:385
221. Sergeeva L, Karabanova L, Grigoryeva O, Zimich O, Gorbach L (1997) Ukr Khim Zh 63:65 (in Russian)
222. Karabanova LV, Boiteux G, Gain O, Seytre G, Sergeeva LM, Lutsyk ED (2001) J Appl Polym Sci 80:852
223. Bartolotta A, Carini G, D'Angelo G, Di Marco G, Farsaci F, Grigoryeva O, Sergeeva L, Slisenko O, Starostenko O, Tripodo G (2004) Philos Mag 84:1591
224. Stepanenko L, Novikova T, Sergeeva L, Shtompel V, Chernobay A (2002) J Appl Chem 75:1341 (in Russian)
225. Tsonos C, Apekis L, Viras K, Stepanenko L, Karabanova L, Sergeeva L (2000) J Macromol Sci Phys B39:155
226. Tsonos C, Apekis L, Viras K, Stepanenko L, Karabanova L, Sergeeva L (2001) Solid State Ionics 143:2299
227. Adache H, Kotaka T (1983) Polym J 15:285
228. Li B, Zhang D, Peng X, Quian B (1983) Polym Commun N 3:202
229. Hourston DJ, Lia Y (1983) J Appl Polym Sci 28:3745
230. Hertmant I, Damyanidu M, Meyer GC (1983) Polymer 24:1419
231. Roh SS, Hong BT, Kim DS (2003) J Appl Polym Sci 87:1079
232. Schneider HA (1989) Polymer 30:77
233. Goldstein M (1985) (1987) Macromolecules 18:277
234. Couchman PR (1987) Polym Eng Sci 27:618
235. Frisch KC, Klempner D, Migdal S, Frisch HL, Giradella H (1974) Polym Eng Sci 14:76
236. Frisch HL, Frisch KC, Klempner D (1974) Polym Eng Sci 14:648
237. Shilov VV, Tsukruk VV, Lipatov YS (1984) Vysokomol Soedin A 26:1347
238. Lipatov YS, Babich VF (1987) Mech Compos Mater 1:17
239. Lipatov YS, Korzhuk NI, Babich VF (1985) In: Molecular characterization of composite interfaces. Plenum, New York, p 237

240. Pentrakoon D (1995) PhD thesis, University of Massachusetts
241. Lipatov YS, Rosovitsky VF (1976) Rep Acad Sci Ukrainian SSR B8:713
242. Rosovitsky VF, Lipatov YS (1986) In: Physical chemistry of multicomponent polymeric systems, vol 2. Naukova Dumka, Kiev, p 229
243. Lipatov YS, Rosovitsky VF, Maslak YV (1984) Vysokomol Soedin A 24:1029
244. Ferry J (1974) Viscoelastic properties of polymers, 2d edn. Wiley, New York
245. Lipatov YS, Rosovitsky VF, Babkina NV (1993) Polymer 34:4697
246. Berstein VA, Egorov VM (1990) Differential scanning calorimetry in physical chemistry of polymers. Chimiya, Leningrad (in Russian)
247. Lindenmeyer PH (1979) J Polym Sci Polym Phys 17:1965
248. Sperling LH, Fay JJ (1991) Polym Adv Technol 2:49
249. Yazdani-Ardakani S, Kesavan SK (1987) Polymer 28:241
250. Chang MCO, Thomas DA, Sperling LH (1987) J Appl Polym Sci 34:409
251. Chang MCO, Thomas DA, Sperling LH (1988) J Polym Sci Polym Phys 26:1627
252. Fay JJ, Thomas DA, Sperling LH (1991) J Appl Polym Sci 43:1623
253. Foster JN, Sperling LH, Thomas DA (1987) J Appl Polym Sci 33:2637
254. Hourston DJ, Zarandouz M (1994) Polyurethane-polydimethylsiloxane simultaneous interpenetrating polymer networks. In: Klempner D, Frisch K (eds) Advances in interpenetrating polymer networks vol 2. Technomic, Lancaster, PA, p 101
255. Rosovitsky VF, Lipatov YS (1985) Dokl Akad Nauk SSSR 283:910
256. Lipatov YS, Rosovitsky VF, Datsko P, Maslak Y (1987) Mech Compos Mater 6:1082 (Riga)
257. Lipatov YS, Rosovitsky VF, Datsko P, Maslak Y (1988) J Appl Polym Sci 36:1143
258. Lipatov YS, Alekseeva TT (1992) Prog Chem 61:2187 (in Russian)
259. Suthar B, Xiao HX, Klempner D, Frisch KC (1996) Polym Adv Technol 7:221
260. Djomo H, Morin A, Damayanidu M, Meyer GC (1983) Polymer 24:65
261. Gladyshev GP, Popov VA (1974) Radical polymerization at high conversion degrees. Nauka, Moscow (in Russian)
262. Meyer GC (1983) Makromol Chem Rapid Commun 4:221
263. Jin SR, Meyer GC (1986) Polymer 227:592
264. Jin SR, Widmaier JM, Meyer GC (1988) Polym Commun 29:26
265. Derrough SN, Rouf C, Widmaier JM, Meyer GC (1991) Polym Mater Sci Eng 65:1
266. Hsu TJ, Lee LJ (1988) J Appl Polym Sci 36:1157
267. Lee LJ (1981) Polym Eng Sci 21:483
268. Yang YS, Lee LJ (1987) Macromolecules 20:1490
269. Ramis X, Cadenato A, Morancho JM, Salla JM (2001) Polymer 42:9469
270. Jenninger W, Schawe JEK, Alig I (2000) Polymer 41:1577
271. Dean K, Cook WD, Burchill P, Zipper M (2001) Polymer 42:3589
272. Ivankovic M, Dzodan N, Brnardic I, Mencer HJ (2002) Appl Polym Sci 83:2689
273. Kulkazni AS, Beaucage G (2005) Polymer 46:4454
274. Jin SR, Widmaier JM, Meyer GC (1988) Polymer 29:347
275. Rouf C, Derrough S, Andre JJ, Widmaier JM, Meyer GC (1994) In: Klempner D, Sperling LH, Utracki LA (eds) Interpenetrating polymer networks. Adv Chem Ser 239. American Chemical Society, Washington, DC, p 143
276. Grigoryeva O, Fainleib A, Pissis P, Boiteux G (2002) Polym Eng Sci 42:2440
277. Nowers JR, Narasimhan B (2006) Polymer 47:1108
278. Cui Y, Chen Y, Wang X, Tian G, Tang X (2002) Polymer 52:1246
279. Fichet O, Vidal F, Laskar J, Teyssie D (2005) Polymer 46:37
280. Laskar J, Vidal F, Fichet O, Gauthier C, Teyssie D (2004) Polymer 45:5047
281. Kiguchi T, Aota H, Matsumoto A (2004) J Appl Polym Sci 94:1198

282. Dean KM, Cook WD (2004) Polym Int 53:1305
283. Widmaier JM, Meyer GC (1989) Chemical aspects in the formation of interpenetrating polymer networks. In: Klempner D, Frisch K (eds) Advances in interpenetrating polymer networks, vol 1. Technomic, Lancaster, PA, p 155
284. Hermant IM, Damayanidu M, Meyer GC (1983) Polymer 24:1419
285. Lipatov YS, Alekseeva TT, Rosovitsky VF (1989) Dokl Akad Nauk SSSR 307:883
286. Kim SK, Kim SC (1990) Polym Bull 23:141
287. Xue SC, Zhang ZP, Wing SK (1989) Polymer 30:1269
288. Semenovich GM, Lipatov YS, Skiba SL, Sergeeva LM (1988) Dokl Akad Nauk SSSR 301:384
289. Lipatov YS, Semerovich GM, Skiba SI, Karabanova LV, Sergeeva LM (1992) Polymer 33:361
290. Lipatov YS, Grigoryeva O (1983) Rep Acad Sci Ukrainian SSR B11:42
291. Shilov VV, Lipatov YS, Karabanova LV, Sergeeva LM (1979) J Polym Sci Polym Chem 17:3083
292. Lipatov YS, Alekseeva TT (1991) Polym Commun 32:254
293. He X, Widmaier JM, Meyer GC (1991) Polym Mater Sci Eng 65:44
294. Tabka N, Widmaier JM, Meyer GC (1991) Plastics Rubber Compos Proc Appl 16:11
295. Zhou P, Xu Q, Frisch HL (1994) Macromolecules 27:938
296. Riccardi CC, Boprrajop J, Meynse L, Fenouillot F, Pascault JP (2004) J Polym Sci Phys 42:1351
297. Lipatov YS, Alekseeva TT, Rosovitsky VF, Babkina NV (1993) Polym Sci A 35:652
298. Stanford JL, Ryan AJ, Yang Y (2001) Polym Int 50:1035
299. Partizel N, Meyer G, Weill G (1993) Polymer 34:2495
300. Nair CPR, Francis T (1999) J Appl Polym Sci 74:3365
301. Sanchez MS, Ferrer GG, Gabanilles CT, Duenas JMM, Pradas MM, Ribelles JLG (2001) Polymer 42:10071
302. Tokareva NN, Duflot VR (1990) Vysokomol Soedin A 32:1250
303. Bol'bit NM, Duflot VR (2002) Polym Sci A 44:394
304. Widmaier JM, Nilly A, Chenal JM, Mathis A (2005) Polymer 46:3318
305. Chenal JM, Widmaier JM (2005) Polymer 46:671
306. Chenal JM, Colombini D, Widmaier JM (2006) J Appl Polym Sci 99:2989
307. Lipatov Y (2007) Polym Bull 58:105
308. Lipatov Y, Kosyanchuk LF, Nesterov AE, Antonenko OI (2003) Polym Int 52:664
309. Alekseeva TT, Yarova NV, Lipatov YS (2005) Rep NASU 3:128 (in Russian)
310. Alekseeva TT, Lipatov YS, Yarova NV (2005) Polym Sci A 47:891
311. Alekseeva TT, Lipatov YS, Babkina NV, Grishchuk SI, Jarova NV (2005) Polymer 46:419
312. Alekseeva TT, Grishchuk SI, Lipatov YS, Babkina NV, Jarova NV (2003) Polym Sci A 45:1237
313. Lipatov YS, Kosyanchuk LF, Jarova NV (2003) Rep Acad Sci Ukraine 2:151
314. Avrami M (1939) J Chem Phys 7:1103
315. Erofeev BV (1949) Dokl Akad Nauk SSSR 52:515
316. Kotova AA, Lanzov VM, Abdrakhmanova LA, Mezhikovsky SM (1987) Vysokomol Soedin A 29:1761
317. Roshchupkin VP, Ozerkovsky BV, Karapetyan ZA (1977) Vysokomol Soedin A 19:2239
318. Alekseeva TT, Lipatova TE, Lipatov YS (1995) Polym Sci A 37:1194
319. Lipatov YS, Alekseeva TT, Shumsky VF (1991) Dokl Akad Nauk SSSR 318:590
320. Zhu S, Tiem Y, Hamillec A, Eaton D (1990) Polymer 31:154

321. Lipatov YS, Alekseeva TT, Shumsky VF (1992) Dokl Akad Nauk UkrSSR B6:131
322. Vinogradov GV, Malkin AY, Plotnikova EP (1978) Vysokomol Soedin A 20:226
323. Mikhal'chuk VM, Stroganoff VF, Lipatov YS (1995) Polym Sci A 37:1672
324. Lipatov YS (1987) Dokl Akad Nauk SSSR 296:646
325. Polak LS (1983) Self-organization in nonequilibrium physico-chemical systems. Nauka, Moscow (in Russian)
326. Lipatov YS (2002) Prog Polym Sci 27:1721
327. Lipatov YS, Alekseeva TT, Rosovitsky VF, Babkina NV (1994) Polym Network Blend 4:9
328. Lipatov YS, Alekseeva TT, Rosovitsky VF, Babkina NV (1995) Polym Int 37:97
329. Seminovych GM, Fainleib AM, Slinchenko EA, Brovko AA, Sergeeva LM, Dudkova VI (1999) React Funct Polym 40:281
330. Brovko AA, Fainleib AM, Slinchenko EA, Dudkova VI, Sergeeva LM (2001) Compos Polym Mater 23:85
331. Paul DR, Newman S (1978) Polymer blends, vol 2. Academic, New York
332. Utracki L (1989) Polymer alloys and blends. Hanser, New York
333. Hermes HE, Higgins JS (1998) Polym Sci Eng 38:847
334. Lipatov YS (1991) Sci Eng Compos Mater 2:69
335. Riese G, Hurtrez G, Bahadier P (1986) Encyclopedia of polymer science and engineering, 2nd edn, vol 2. Wiley, New York, p 390
336. Nevissas V, Widmaier JM, Mayer GC (1988) J Appl Polym Sci 36:1467
337. Hourston DJ, Zia Y (1984) J Appl Polym Sci 29:629
338. Scarito PR, Sperling LH (1974) J Polym Sci C 46:175
339. Hourston DJ, Schafer FU (1997) In: Kim SC, Sperling LH (eds) IPNs around the world: science and engineering. Wiley, New York, p 155
340. Hourston DJ, Schafer FU (1996) Polym Adv Technol 7:273
341. Lipatov YS, Alekseeva TT, Babkina NV (2001) J Polym Mater 18:201
342. Cassidy EF, Xiao HX, Frisch KC, Frisch HL (1984) J Pol Sci Polym Chem Ed 22:1851
343. Hsieh KH, Chou LM, Chiang YC (1989) Polym J 21:1
344. Han X, Chen B, Guo F (1997) In: Kim SC, Sperling LH (eds) IPNs around the world: science and engineering. Wiley, New York, p 241
345. Yu X, Gao G, Wang J, Li F, Tang X (1999) Polym Int 48:805
346. Wang MV, Lee CT, Lin MS (1997) Polym Int 44:503
347. He X, Widmaier JM, Meyer GC (1993) Polym Int 32:289
348. Lipatov YS, Alekseeva TT (1997) In: Kim SC, Sperling LH (eds) IPNs around the world: science and engineering. Wiley, New York, p 75
349. Suthar B, Xiao HX, Klempner D, Frisch KC (1997) In: Kim SC, Sperling LH (eds) IPNs around the world: science and engineering. Wiley, New York, p 49
350. Kim JH, Kim SC (1987) Polym Eng Sci 27:1243
351. Kim JH, Kim SC (1987) Polym Eng Sci 27:1252
352. Alekseeva TT, Lipatov YS, Babkina NV (2001) Macromol Symp 164:91
353. Alekseeva TT, Lipatov YS, Grishchuk SI, Babkina NV (2003) Macromol Symp 210:219
354. Babkina NV, Lipatov YS, Alekseeva TT (2005) Polym Sci A 47:2118
355. Eklind H (1996) Interphases in polymer blends. Thesis, Chalmers University of Technology
356. Beckman EJ, Karasz FE, Porter RS, MacKnight WJ (1988) Macromolecules 21:1193

Editor: Karel Dušek

Author Index Volumes 201–209

Author Index Volumes 1–100 see Volume 100
Author Index Volumes 101–200 see Volume 200

Alekseeva, T., see Lipatov, Y. S.: Vol. 208, pp. 1–227
Anwander, R. see Fischbach, A.: Vol. 204, pp. 155–290.
Ayres, L. see Löwik D. W. P. M.: Vol. 202, pp. 19–52.

Binder, W. H. and *Zirbs, R.*: Supramolecular Polymers and Networks with Hydrogen Bonds in the Main- and Side-Chain. Vol. 207, pp. 1–78
Bouteiller, L.: Assembly via Hydrogen Bonds of Low Molar Mass Compounds into Supramolecular Polymers. Vol. 207, pp. 79–112
Boutevin, B., David, G. and *Boyer, C.*: Telechelic Oligomers and Macromonomers by Radical Techniques. Vol. 206, pp. 31–135
Boyer, C., see Boutevin B: Vol. 206, pp. 31–135
ten Brinke, G., Ruokolainen, J. and *Ikkala, O.*: Supramolecular Materials Based On Hydrogen-Bonded Polymers. Vol. 207, pp. 113–177

Csetneki, I., see Filipcsei G: Vol. 206, pp. 137–189

David, G., see Boutevin B: Vol. 206, pp. 31–135
Deming T. J.: Polypeptide and Polypeptide Hybrid Copolymer Synthesis via NCA Polymerization. Vol. 202, pp. 1–18.
Dong Liu, X., Yamada, M., Matsunaga, M. and *Nishi, N.*: Functional Materials Derived from DNA. Vol. 209, pp. 149–178
Donnio, B. and *Guillon, D.*: Liquid Crystalline Dendrimers and Polypedes. Vol. 201, pp. 45–156.

Elisseeff, J. H. see Varghese, S.: Vol. 203, pp. 95–144.
Esker, A. R., Kim, C. and *Yu, H.*: Polymer Monolayer Dynamics. Vol. 209, pp. 59–110

Ferguson, J. S., see Gong B: Vol. 206, pp. 1–29
Filipcsei, G., Csetneki, I., Szilágyi, A. and *Zrínyi, M.*: Magnetic Field-Responsive Smart Polymer Composites. Vol. 206, pp. 137–189
Fischbach, A. and *Anwander, R.*: Rare-Earth Metals and Aluminum Getting Close in Ziegler-type Organometallics. Vol. 204, pp. 155–290.
Fischbach, C. and *Mooney, D. J.*: Polymeric Systems for Bioinspired Delivery of Angiogenic Molecules. Vol. 203, pp. 191–222.
Freier T.: Biopolyesters in Tissue Engineering Applications. Vol. 203, pp. 1–62.
Friebe, L., Nuyken, O. and *Obrecht, W.*: Neodymium Based Ziegler/Natta Catalysts and their Application in Diene Polymerization. Vol. 204, pp. 1–154.

García A. J.: Interfaces to Control Cell-Biomaterial Adhesive Interactions. Vol. 203, pp. 171–190.

Gong, B., Sanford, AR. and *Ferguson, JS.*: Enforced Folding of Unnatural Oligomers: Creating Hollow Helices with Nanosized Pores. Vol. 206, pp. 1–29

Guillon, D. see Donnio, B.: Vol. 201, pp. 45–156.

Harada, A., Hashidzume, A. and *Takashima, Y.*: Cyclodextrin-Based Supramolecular Polymers. Vol. 201, pp. 1–44.

Hashidzume, A. see Harada, A.: Vol. 201, pp. 1–44.

Häußler, M. and *Tang, B. Z.*: Functional Hyperbranched Macromolecules Constructed from Acetylenic Triple-Bond Building Blocks. Vol. 209, pp. 1–58

Heinze, T., Liebert, T., Heublein, B. and *Hornig, S.*: Functional Polymers Based on Dextran. Vol. 205, pp. 199–291.

Heßler, N. see Klemm, D.: Vol. 205, pp. 57–104.

Van Hest J. C. M. see Löwik D. W. P. M.: Vol. 202, pp. 19–52.

Heublein, B. see Heinze, T.: Vol. 205, pp. 199–291.

Hornig, S. see Heinze, T.: Vol. 205, pp. 199–291.

Hornung, M. see Klemm, D.: Vol. 205, pp. 57–104.

Ikkala, O., see ten Brinke, G.: Vol. 207, pp. 113–177

Jaeger, W. see Kudaibergenov, S.: Vol. 201, pp. 157–224.

Janowski, B. see Pielichowski, K.: Vol. 201, pp. 225–296.

Kataoka, K. see Osada, K.: Vol. 202, pp. 113–154.

Kim, C., see Esker, A. R.: Vol. 209, pp. 59–110

Klemm, D., Schumann, D., Kramer, F., Heßler, N., Hornung, M., Schmauder H.-P. and *Marsch, S.*: Nanocelluloses as Innovative Polymers in Research and Application. Vol. 205, pp. 57–104.

Klok H.-A. and *Lecommandoux, S.*: Solid-State Structure, Organization and Properties of Peptide—Synthetic Hybrid Block Copolymers. Vol. 202, pp. 75–112.

Kosma, P. see Potthast, A.: Vol. 205, pp. 151–198.

Kosma, P. see Rosenau, T.: Vol. 205, pp. 105–149.

Kramer, F. see Klemm, D.: Vol. 205, pp. 57–104.

Kudaibergenov, S., Jaeger, W. and *Laschewsky, A.*: Polymeric Betaines: Synthesis, Characterization, and Application. Vol. 201, pp. 157–224.

Laschewsky, A. see Kudaibergenov, S.: Vol. 201, pp. 157–224.

Lecommandoux, S. see Klok H.-A.: Vol. 202, pp. 75–112.

Li, S., see Li W: Vol. 206, pp. 191–210

Li, W. and *Li, S.*: Molecular Imprinting: A Versatile Tool for Separation, Sensors and Catalysis. Vol. 206, pp. 191–210

Liebert, T. see Heinze, T.: Vol. 205, pp. 199–291.

Lipatov, Y. S. and *Alekseeva, T.*: Phase-Separated Interpenetrating Polymer Networks. Vol. 208, pp. 1–227

Löwik, D. W. P. M., Ayres, L., Smeenk, J. M., Van Hest J. C. M.: Synthesis of Bio-Inspired Hybrid Polymers Using Peptide Synthesis and Protein Engineering. Vol. 202, pp. 19–52.

Lucas, P. and *Robin, J.-J.*: Silicone-Based Polymer Blends: An Overview of the Materials and Processes. Vol. 209, pp. 111–147

Marsch, S. see Klemm, D.: Vol. 205, pp. 57–104.
Matsunaga, M., see Dong Liu, X.: Vol. 209, pp. 149–178
Mooney, D. J. see Fischbach, C.: Vol. 203, pp. 191–222.

Nishi, N., see Dong Liu, X.: Vol. 209, pp. 149–178
Nishio Y.: Material Functionalization of Cellulose and Related Polysaccharides via Diverse Microcompositions. Vol. 205, pp. 1–55.
Njuguna, J. see Pielichowski, K.: Vol. 201, pp. 225–296.
Nuyken, O. see Friebe, L.: Vol. 204, pp. 1–154.

Obrecht, W. see Friebe, L.: Vol. 204, pp. 1–154.
Osada, K. and *Kataoka, K.*: Drug and Gene Delivery Based on Supramolecular Assembly of PEG-Polypeptide Hybrid Block Copolymers. Vol. 202, pp. 113–154.

Pielichowski, J. see Pielichowski, K.: Vol. 201, pp. 225–296.
Pielichowski, K., Njuguna, J., Janowski, B. and *Pielichowski, J.*: Polyhedral Oligomeric Silsesquioxanes (POSS)-Containing Nanohybrid Polymers. Vol. 201, pp. 225–296.
Pompe, T. see Werner, C.: Vol. 203, pp. 63–94.
Potthast, A., Rosenau, T. and *Kosma, P.*: Analysis of Oxidized Functionalities in Cellulose. Vol. 205, pp. 151–198.
Potthast, A. see Rosenau, T.: Vol. 205, pp. 105–149.

Robin, J.-J., see Lucas, P.: Vol. 209, pp. 111–147
Rosenau, T., Potthast, A. and *Kosma, P.*: Trapping of Reactive Intermediates to Study Reaction Mechanisms in Cellulose Chemistry. Vol. 205, pp. 105–149.
Rosenau, T. see Potthast, A.: Vol. 205, pp. 151–198.
Rotello, V. M., see Xu, H.: Vol. 207, pp. 179–198
Ruokolainen, J., see ten Brinke, G.: Vol. 207, pp. 113–177

Salchert, K. see Werner, C.: Vol. 203, pp. 63–94.
Sanford, A. R., see Gong B: Vol. 206, pp. 1–29
Schlaad H.: Solution Properties of Polypeptide-based Copolymers. Vol. 202, pp. 53–74.
Schmauder H.-P. see Klemm, D.: Vol. 205, pp. 57–104.
Schumann, D. see Klemm, D.: Vol. 205, pp. 57–104.
Smeenk, J. M. see Löwik D. W. P. M.: Vol. 202, pp. 19–52.
Srivastava, S., see Xu, H.: Vol. 207, pp. 179–198
Szilágyi, A., see Filipcsei G: Vol. 206, pp. 137–189

Takashima, Y. see Harada, A.: Vol. 201, pp. 1–44.
Tang, B. Z., see Häußler, M.: Vol. 209, pp. 1–58

Varghese, S. and *Elisseeff, J. H.*: Hydrogels for Musculoskeletal Tissue Engineering. Vol. 203, pp. 95–144.

Wang, D.-A.: Engineering Blood-Contact Biomaterials by "H-Bond Grafting" Surface Modification. Vol. 209, pp. 179–227
Werner, C., Pompe, T. and *Salchert, K.*: Modulating Extracellular Matrix at Interfaces of Polymeric Materials. Vol. 203, pp. 63–94.

Xu, H., Srivastava, S., and *Rotello, V. M.*: Nanocomposites Based on Hydrogen Bonds. Vol. 207, pp. 179–198

Yamada, M., see Dong Liu, X.: Vol. 209, pp. 149–178
Yu, H., see Esker, A. R.: Vol. 209, pp. 59–110

Zhang, S. see Zhao, X.: Vol. 203, pp. 145–170.
Zhao, X. and *Zhang, S.*: Self-Assembling Nanopeptides Become a New Type of Biomaterial. Vol. 203, pp. 145–170.
Zirbs, R., see Binder, W. H.: Vol. 207, pp. 1–78
Zrínyi, M., see Filipcsei G: Vol. 206, pp. 137–189

Subject Index

AIPNs   120
Amorphous polymer networks, formation/structure   10
Anamorphoses   164
Anionic polymerization, sequential IPNs   80
Avrami–Erofeev equation   183

Benzene   22
Binder–Frisch approach   44
Bismaleimide–triazine   179
Bisphenol A based bismaleimide (BMP)   179
Bisphenol A based cyanate ester (BACY)   179

Chain diffusion, reptation mechanism   14
Chemical kinetics, IPN formation/phase separation   147
Chemical quenching   44
Coalescence   28
Compatibilization, phase-separated IPNs   199
Compatibilizers   207, 215
Constituent networks, kinetics of IPN formation   157
Copolymer–benzene   22
Cross-linking density   56
Curing conditions   116
Cyanate ester resins (CERs)   120

Divinylbenzene (DVB)   16, 74
DMA investigation   210
Domain formation   25
Domain sizes   128
DSC measurements   214
Dual-phase continuity   15
Dynamic mechanical analysis (DMA)   179
Dynamic shear modulus   132

Entanglements   14
Epoxy–diene   129
Evolved phases, kinetics   176

Filler, kinetics   194
Forced compatibility   15
Free energy of mixing   30
Free volume   62
FTIR   179

Gel formation time   166
Gelation   183
Gibbs free energy   21
Glass transition temperatures   128
Gradient IPN–benzene   22
Growth   24
Guest polymer   9

Heat capacity changes   175
Heterogeneity   70
–, experimental data   73
Heterogeneous structure   55
–, parameters   70
Host polymer   9
Hybrid binders   14

Immiscibility   21
Interdiffusion coefficients   40
Interfacial region   47
Internetwork grafting   204
Interphase   214
–, role of kinetics   181
–, viscoelastic properties   126
IPN formation, kinetics   148
IPN mixing, thermodynamics   15
IPN morphology   96

Kinetics, chemical, IPN formation/phase separation   147

Lower critical solution temperature (LCST)
    14

Macrosyneresis   11
Mechanical models   105
Methyl methacrylate (MMA)   19
Microgel   183
Microphase separation   34
Microphase structure   87
Microsyneresis   11
Miscibility   21
MMA–PMMA
Morphology   55

Network theory, rubber elasticity   13
Nonequilibrium states   52
Nucleation   24

Oligoesteracrylate   161

PBMA, semi-IPN   164
PEO/PMMA   20
Phase diagrams, reacting systems   16
Phase separation   12, 15, 24
–, composition/ratio of phases   51
–, thermodynamics/kinetics   173
Phenyl butyl acetate (PnBA)   28
PMA/PMMA   179
Polybutadiene, cross-linked   20
Polycatenanes   8
Polymeric catenanes   8
Poly-α-methylstyrene, cross-linked   21
Polyurethane (PU), cross-linked   17
–, butyl methacrylate   169
–, crystallizable   120
–, linear segmented (LPUs)   120
–, simultaneous IPNs   77
Poly(acrylic acid)   58
Poly(butyl methacrylate) (PBMA)   16
Poly(ethylene oxide)   20
Poly(methyl methacrylate) (PMMA)   19
Poly(oxyethylene)   58
Poly(oxypropylene glycol) (POPG)   61
Poly(urethane acrylate) (PUA)   22, 77
Poly(vinyl acetate) (PVA), linear   17
Poly(vinyl methyl ether) (PVME)   42
PS, linear   21
PU network, PBMA, semi-IPN   167
PU/benzene   22
PU/ionomeric PU, microphase structure
    84

PU/PBMA   172
–, compatibilization   203
PU/PS, compatibilization   208
PU/PVC   18
PVA–PBMA–PU   17

Reaction conditions   87
Reaction kinetics, microphase separation
    164
–, phase transition   183
Relaxation spectra   130
Relaxation transitions   103, 111
Reptation mechanism, chain diffusion   14
Rheokinetics, semi-IPNs, formation   186
Rotaxanes   8

Segregation   181
Segregation degree   143
–, relaxation maxima   139
Self-organization   191
Sequential IPNs   9
–, kinetics of formation   179
–, polyurethane   74
–, styrene–divinylbenzene copolymer   74
Simultaneous IPNs   9
Small-angle neutron scattering (SANS)   91
Spinodal decomposition   35
Stocks–Einstein equation   32
Structure states, semi-IPNs, formation
    186
Styrene–divinylbenzene copolymer   74
Styrene–DVB copolymer, semi-IPN   164
Styrene–DVB–PBMA   16

Temperature transitions   124
Thermodynamics   15
Thermoelastoplastics   11
Thermogravimetric analysis (TGA)   179
Thermoplastic apparent IPNs   120
Topological isomerism   7
Two-phase microheterogeneous systems
    15

Upper critical solution temperature (UCST)
    14

Vacancies   67
Vibration-damping   137
Viscoelastic functions   111
Viscoelastic properties, segregation degree
    143
Viscoelasticity   103, 116

Printing: Krips bv, Meppel, The Netherlands
Binding: Stürtz, Würzburg, Germany